Advances in Neurotraumatology

Published under the Auspices of the Neurotraumatology Committee of the World Federation of Neurosurgical Societies

Editor-in-Chief: R. P. Vigouroux

Volume 3

Cerebral Contusions, Lacerations and Hematomas

Managing Editor: R. A. Frowein

Contributors:

R. Firsching
G. Foroglou
G. Friedmann
R. A. Frowein
J. W. Glowacki
P. Guillermain
N. Nakamura

I. Oprescu†
P. Rabehanta
K. E. Richard
D. A. Stålhammar
U. Stammler
F. Thun
R. P. Vigouroux

Language Editor: Ph. Harris

Springer-Verlag Wien New York

Professor ROBERT P. VIGOUROUX
Clinique Neuro-Chirurgicale, C.H.U. Timone, Marseille, France

Professor REINHOLD A. FROWEIN
Neurochirurgische Universitätsklinik Köln, Federal Republic of Germany

© 1991 by Springer-Verlag/Wien
Softcover reprint of the hardcover 1st edition 1991
Typeset by Macmillan India Ltd., Bangalore 25

With 136 Figures

ISSN 0178-3696
ISBN-13: 978-3-7091-7435-7   e-ISBN-13: 978-3-7091-6922-3
DOI: 10.1007/978-3-7091-6922-3

Foreword

Posttraumatic cerebral contusions, lacerations and hematomas in the past could often only be suspected by clinical symptomatology and be visualized by angiography in a restricted manner, but they are now diagnosed with precision through CT and MRI; they remain in the limelight in our daily management of severe head injuries.

Stålhammar's longlasting research in biomechanics is here condensed in a concise review of the current knowledge in this field, thus providing the basis for our understanding of the parenchymal cerebral posttraumatic lesions.

The neuropathological investigations and findings remain fundamental to the clinical features, and in this text there is the advantage that they were carried out and interpreted by a very experienced neurosurgeon, Ion Oprescu, who most regretfully died before the completion of this volume.

The special morphological and clinical investigations by Nakamura, concerning diffuse brain injury, contribute to our necessary diagnosis and treatment of this phenomenon where, without gross visible lesions of cerebral tissue, a high mortality occurs.

The clinical features, the principles of therapy and the outcome are described by Vigouroux and Guillermain, whose clinical study clearly demonstrates the limitations of all our efforts in severe brain injuries, in spite of decades of clinical experience and research.

The analysis of the different combinations of brain swelling, edema, contusions and hematomas by Richard illustrates the interaction of these pathological lesions and their important time relationships, such as early brain swelling and the later appearance of brain edema, as well as the correlation with the age of the patients in respect to the outcome.

The clinical and radiological manifestations of intracerebral hematomas are examined by Foroglou, and various theories concerning their etiology are discussed, some of which are obviously based on too restricted numbers of patients. The nomenclature is colourful and it cannot be denied that even the authors of this volume do not use uniform terms.

Based on a particularly broad experience with posttraumatic hematomas, Glowacki presents the changes in the management of cerebellar contusions and hematomas after the advent of modern imaging methods.

Frowein and coworkers underline again, from the clinical and radio-logical points of view, the important factor of time not only for the recognition of contusional lesions, but also for the reliability of diagnosis. From their study it becomes clear why the prognosis becomes more realistic no sooner than 24 hours after trauma. The optimum time of encountering a hematoma is within the first three hours; and of an enlarging contusion it is 8 to 12 hours following injury. Their analysis provides a comprehensive definition of the dynamics of posttraumatic contusions and of intracerebral hematomas.

The growing interest in multimodality evoked potentials in the post-traumatic period is shown in Firsching's current review of the value of these special neurophysiological investigations.

The editors sincerely wish to thank all of the authors for their efforts and their valid contributions. They are particularly grateful to Phillip Harris for his irreplaceable and constructive advice as language editor. We also thank the publishers for their patient and continuous help to reach our common goal, to provide realistic basic data for the best possible care for our severely head injured patients.

And we wish to thank Mrs Lieselotte Jahn for her valuable secretary work.

ROBERT P. VIGOUROUX   REINHOLD A. FROWEIN

Contents

List of Contributors

Firsching, P. D. Dr. R., Neurochirurgische Universitäts-Klinik, Joseph Stelzmann Strasse 9, D-W-5000 Köln 41, Federal Republic of Germany.

Foroglou, Professor Dr. G., Director, Department of Neurological Surgery, "Ahepa" General Hospital University of Thessaloniki, Thessaloniki, Greece.

Friedmann, Professor Dr. G., Radiologisches Institut der Universität, Joseph Stelzmann Strasse 9, D-W-5000 Köln 41, Federal Republic of Germany.

Frowein, Professor Dr. R. A., em. Direktor der Neurochirurgischen Universitäts-Klinik zu Köln, Geibel Strasse 22, D-W-5000 Köln 41, Federal Republic of Germany.

Glowacki, Professor Dr. J. W., Klinika Neurotraumatologii AM, Botaniczna 3, PL-31-503 Krakow, Poland.

Guillermain, Professor Dr. P., Clinique Neuro-Chirurgicale, C.H.U. Timone, 264, rue Saint-Pierre, F-13385 Marseille Cedex 5, France.

Harris, Professor Dr. P., Department of Surgical Neurology, Western General Hospital, Crewe Road, Edinburgh EH4 2XU, U.K.

Nakamura, Professor Dr. N., Department of Neurosurgery, The Jikei University School of Medicine, 3-25-8, Nishi-Shimbashi, Minato-Ku, Tokyo, Japan.

Rabehanta, Professor Dr. P., Clinique Neuro-Chirurgicale, C.H.U. Timone, 264, rue Saint-Pierre, F-13385 Marseille Cedex 5, France.

Richard, Professor Dr. K. E., Neurochirurgische Universitäts-Klinik, Joseph Stelzmann Strasse 9, D-W-5000 Köln 41, Federal Republic of Germany.

Stålhammar, Professor Dr. D. A., Göteborgs Universitet Neurokirurgiska institutionen, Sahlgrenska sjukhuset, S-41345 Göteborg, Sweden.

Stammler, Dr. U., Neurochirurgische Universitäts-Klinik, Joseph Stelzmann Strasse 9, D-W-5000 Köln 41, Federal Republic of Germany.

Thun, Dr. F., Radiologisches Institut der Universität, Joseph Stelzmann Strasse 9, D-W-5000 Köln 41, Federal Republic of Germany.

Vigouroux, Professor Dr. R. P., Clinique Neuro-Chirurgicale, C.H.U. Timone, 264, rue Saint-Pierre, F-13385 Marseille Cedex 5, France.

Biomechanics of Brain Injuries

D.A. Stålhammar

Department of Neurosurgery, Sahlgren's Hospital, Göteborg (Sweden)

With 10 Figures

Contents

Introduction

The biomechanics of brain injuries describes how the impact of forces directed to the head are related to the primary physical and physiological effects in the skull and the brain.

This chapter discusses the general principles of injury mechanics regarding tissue deformation and its temporal characteristics. Some physical and anatomical features are presented as a background to the main theories of brain injury mechanisms. Biomechanically brain injuries can be classified as those mainly caused by contact forces, and/or inertial forces, and from a pathoanatomical point of view a distinction is made between focal and diffuse injuries. The relation between these categories of injury and the components of loading is explained. Since prevention of injuries is an essential part of trauma management the application of biomechanics for the reduction of injuries is summarized.

Basic Injury Mechanisms

Force, Strain, Duration of Load

Brain injury, *i.e.* dysfunction and structural failure as a result of mechanical load (force, stress), is caused by the *relative motions* (*strains*) generated within the tissues (Goldsmith 1966, von Gierke 1966).

When a *force* is applied to an animate structure this will be deformed and accelerated; the available energy and the area of contact being the primary determinants for the mechanical and physiological responses.

If the body is undeformable, its acceleration is directly related to the applied force according to Newtons' second law (Force = acceleration × mass). At direct impact and impulsive loading (*cf.* Fig. 4), the applied force varies rapidly over time and lasts from a few milliseconds up to about 200 ms (Holbourn 1943), usually in the range of 5–50 ms. The mechanical effects in the tissues are related to time derivatives and gradients of the deformation, *i.e.* displacement, velocity, acceleration and the rate of acceleration change.

Injurious Parameters

Depending on the *ratio* between the *duration of load* and the *natural period of the loaded body* (the duration of one oscillation at free vibration), different characteristics of the acceleration can be identified as the *critical injurious parameters*: its average or peak magnitude, its velocity change (*i.e.* the average acceleration times the pulse duration) or both. Although the basic parameters are acceleration and duration (*cf.* Figs. 9 and 10) their relationship is commonly presented as a sensitivity curve (Fig. 1), composed of the velocity change, the average acceleration and the ratio between the applied pulse and the natural period for the system (Kornhauser 1954).

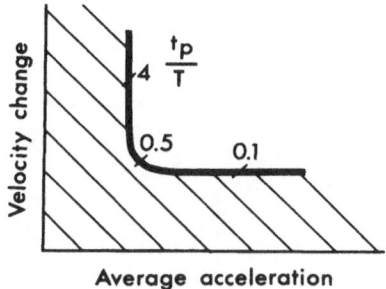

Fig. 1. Inertia–sensitivity curve for a mass-spring system subjected to rectangular pulses. t_p duration of pulse, T natural period of the system. The hatched area is safe. At short load durations an increased velocity change produces damage, while at long durations, damage depends on the magnitude of accelerations. (Modified from Kornhauser 1954 and reprinted from Stålhammar[53] by permission)

If the ratio is small compared to one (*i.e.* short duration loads) the same *velocity change* (impulse) will result in the same injury even if the acceleration (force) varies over a wide range. The blow is over before the structure has begun to be distorted. For long duration loads the same *acceleration* (force) will produce a similar injury even if the velocity change and duration varies over a wide range. If the ratio is about one half, neither the velocity change (impulse) nor the acceleration (force) alone can characterize the sensitivity and consequently the complete *acceleration–time* (force–time) *history* must be considered (von Gierke 1966).

Since the natural frequency of the human brain has been estimated to be about 30 Hz, the duration of loading in real life trauma and in experimental situations (5–50 ms) is within the time interval where acceleration, as well as velocity change have to be considered (*cf.* Fig. 1).

It should be noted that complex structures, like living tissues, present several modes of failure (Fig. 2) (Ljung *et al.* 1983, *cf.* Kornhauser and Gold 1962).

Specific Brain Injury Mechanisms

The tissue strains, being the ultimate direct cause of brain injury, can be generated by different mechanisms: (Fig. 3): skull deformation, angular acceleration, the pressure gradient – translational acceleration and movements in the cranio-spinal junction.

The characteristics of the head structures and of the different types of loading and their distinct effects on the skull and the brain have bearing upon the mechanisms. Here only blunt direct head impacts, predominant in civilian accidents will be discussed.

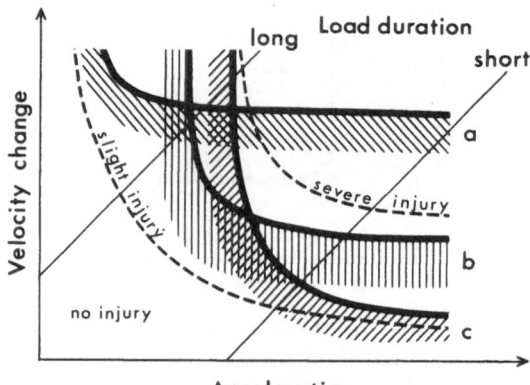

Fig. 2. Sensitivity curves for three hypothetical components (*a*, *b*, and *c*) of a structure (hatched areas indicate ranges of functional failure), and for the resulting overall slight and severe injury. *Cf.* Fig. 10. (Modified from Ljung *et al.* 1983 and reprinted from Stålhammar[53] by permission)

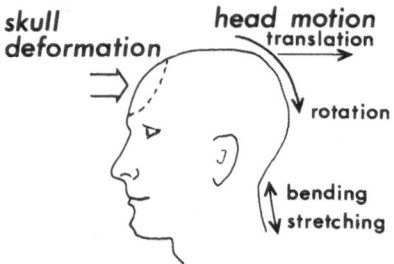

Fig. 3. Impact to the skull produces skull deformation and head motion. The latter comprising three components; translation, rotation and bending–stretching in the cranio-spinal junction. (Reprinted from Stålhammar[53] by permission)

Anatomical Features

The viscoelastic *scalp* is of importance for distribution of the load and the prevention of sharp load peaks. The scalp may reduce the available impact energy at a typical blunt direct head impact up to 13% (Melvin and Evans 1971), at a glancing blow the transferred energy may be almost negligible.

The *skull*, the hard shell enclosing the brain, can be regarded as a combination of a hemisphere and a sphere, which is of importance for the pressure response at impact (Lindgren 1966). The falx and the tentorium are important for brain movements (Aldman *et al.* 1982). This particular shape of the neurocranium, together with its mobile attachment to the neck, will modify the response of the intracranial contents differently depending on the site and direction of impact (Lindgren 1983).

The sutures of the neurocranium close completely with bone between 20–40 years of age. This progressive closure of the skull bones changes the mechanical characteristics of the skull and probably contributes to a different mechanical response of the skull and its contents to impact, in young adults compared to the adult. Löfgren and Pelletieri (1967) advanced such circumstances to explain the lower incidence to temporal lobe contusions at occipital impact below the age of thirty.

Further features of the skull determining the injury mechanics are the shape of the smooth inner surface, in the upper part of the skull and highly irregular in the base. Large areas of the skull base are very thin (*e.g.* the orbital plates) and concentration of stress there will cause deformation, fractures and injuries of the brain surface. The characteristic structure of the skull bone with an inner and outer table of compact bone separated by a layer of spongy bone in combination with the dome-like shape provides excellent protection even against forceful mechanical loading.

Static loading of 5000–7000 N of fresh skulls will cause a fracture, while 4–5 ms dynamic loading may produce a linear fracture at a peak force of 4500–10000 N (Hodgson 1967).

The brain can be considered as a viscoelastic structure soft and yielding "lacking both the rigidity usually associated with a gel, and the plasticity of a paste" (Dodgson 1962 cited by Lindgren 1966).

Comprehensive information on the mechanical properties of the brain and related structures is available in: Lindgren 1966, Ommaya 1968, Liu 1979, and Thibault and Gennarelli 1985.

Characteristics of Loading

In typical accidents the head is injured by an impact lasting about 5–20 ms. An impact generates *contact forces* of high magnitude and short duration, and usually a sudden change of motion which, in turn, generates *inertial forces*. Depending on magnitude, duration and direction of the force, the contact surface, the site of application, the effects of either contact or inertial forces may dominate. A sharp, light object will create changes mainly at the site of impact, while a boxing blow against the face will cause negligible deformation of the neurocranium although the brain injury may be fatal.

The character of the load, particularly its duration, forms a basis for a subdivision of brain injury mechanisms into three main categories; *impacts, impulsive loads* and *static* (compressive) *loads*. The relation between these three types of loading and the injurious mechanical events are shown in Fig. 4; different physical phenomena may produce different functional and structural changes, in separate parts of the brain.

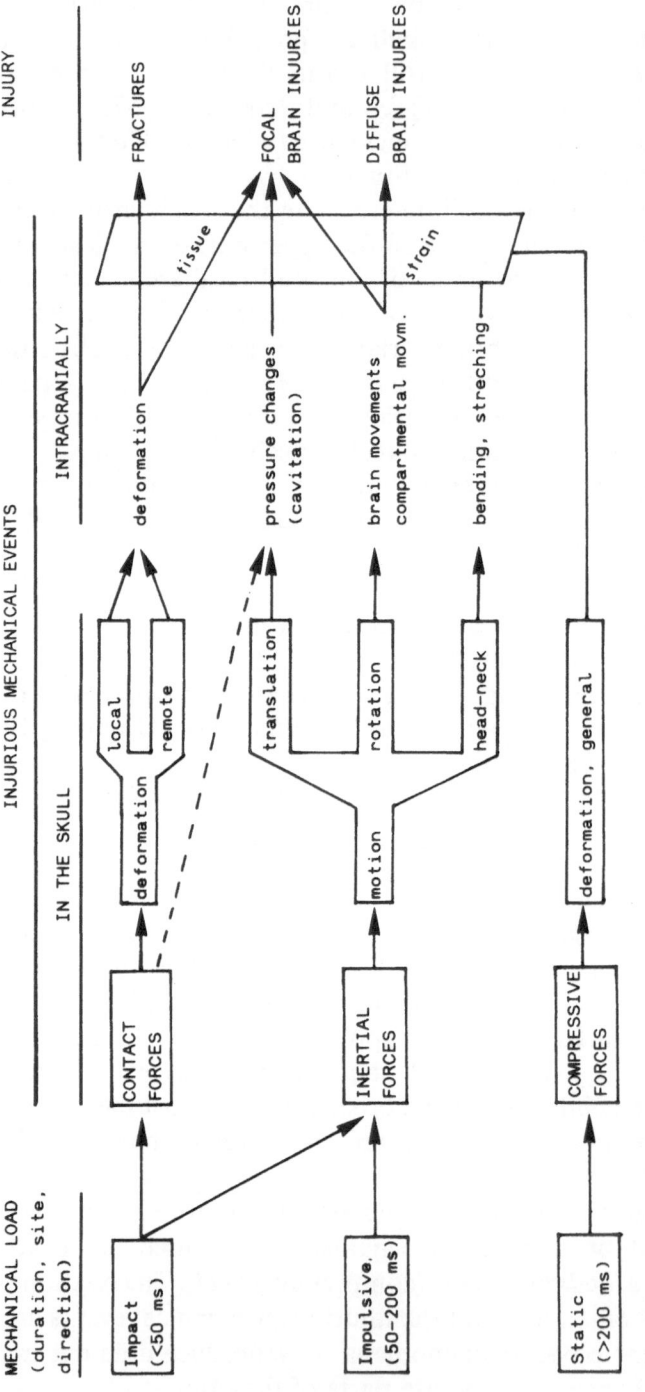

Fig. 4. Relation between injury and mechanical load at head trauma. Potentially injurious mechanical events, regarding the skull and its contents are indicated. Interrelations between these events and in turn, their correlations with characteristics of the mechanical load and category of injury are outlined according to proposed and tested hypothesis. (Modified from Lindgren 1981 and Ommaya 1985 and reprinted from Stålhammar[53] by permission)

Cranial Effects of Loading

The contact forces initiate two distinct mechanical effects in the cranium: stress waves and skull deformation, the latter often described as part of so-called contact phenomena. The *stress waves* are disturbances moving through the structures as "states of strain and particle motion" (Goldsmith 1966). However, the damaging effect of the stress waves has been questioned because their transit time is short (about 0.12 ms) compared to the durations of impacts (usually > 2 ms), and in addition, they are probably attenuated because of the viscous nature of the tissue (Goldsmith 1972). *Skull deformation* occurs at the site of impact and at other locations. Irrespective of whether the deformation is reversible or results in a permanent indentation or fracture and penetration of the bone, it may cause focal injury to the brain and related structures, such as contusions and hemorrhages.

Pure *impulsive loading* implies that effects of contact can be ignored; the sudden change in head motion (acceleration or deceleration) generates brain movements because of inertia. However, as indicated in Fig. 4, head motion may well be caused by an impact. Three components of head motion are discerned; translation, rotation and bending-stretching at the head-neck junction. Translation denotes that the head's centre of gravity moves along a straight line and rotation that the head moves around the centre of gravity. Obviously both these types of motion normally take place simultaneously, and the centre of movement is located outside the centre of gravity, often extracranially. Such a motion may be denoted angular or centroid.

Translational skull motion causes movement of the brain, both absolute and relative to the skull bone and, moreover, creates intracranial pressure changes (Fig. 5). In the area opposite to the contact, negative pressure transients may cause the formation of cavitation bubbles, which at their collapse, may possibly contribute to brain damage. Although these effects

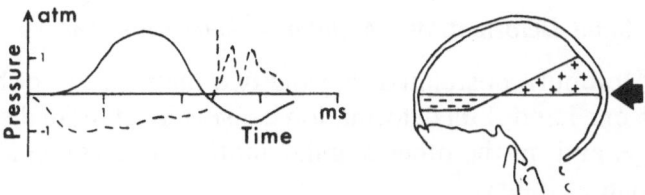

Fig. 5. Occipital impact to a cadaver skull produces initially a positive pressure area (+) beneath the site of impact and a negative pressure area (−) in the frontal region. (Modified from Lindgren 1966, p 89 and reprinted from Stålhammar[53] by permission)

D.A. Stålhammar

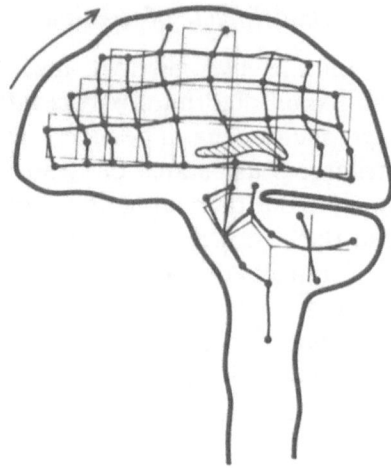

Fig. 6. Deformation of a grid introduced in a silicone gel in a skull model subjected to a sudden angular acceleration in the direction of the arrow. Liquid paraffine substitutes the cerebrospinal fluid in the gel-skull interface. Maximum deformation occurs some distance below that interface. (Redrawn from Aldman *et al.* 1982 and reprinted from Stålhammar[53] by permission)

are mainly related to translational motion of the head, the pressure changes are possibly, in part, elicited by stress waves through the brain, *i.e.* effects of contact forces.

At *rotational head motion* the brain lags behind the skull and stress (inertial forces) is exerted upon the connections between the brain and the skull-dura (bridging veins), and upon the brain tissue itself (Fig. 6). The strains thus elicited can cause rupture of the bridging veins, resulting in subdural hematomas, and of the neural and vascular tissue of the brain parenchyma, resulting in widespread axonal damage and bleeding.

Skull Deformation – Angular Acceleration Theory

In 1943 Holbourn postulated that the two main causes of brain injury would be, on one hand skull deformation causing local brain distortion and brain damage, and on the other angular motion, generating shear strains and diffuse neural injury.

The demonstration of movements of the brain surface at blows to the monkey's head (Pudenz and Shelden 1946) and of widespread alterations in the white matter in autopsy material (Strich 1956) seemed to support the Holbourn hypothesis.

Later Ommaya *et al.* (1964) showed that monkeys, subjected to occipital blows, could be protected from experimental concussion by a cervical collar, presumably because of reduced head rotation. By a new head-accelerating device (HAD-II) Higgins and Schmall (1967) could produce angular accelerations by a 'distributed' impact to the head fitted into a helmet by plaster of Paris. These experiments evidenced that experimental concussion and intracranial hemorrhages (subdural, subarachnoidal and in the superficial cortical layers) were produced by angular acceleration (Unterharnscheidt and Higgins 1969). However there seemed to be a difference between cortical contusion from angular trauma compared to those from pure translational motion. Similarly Ommaya and Gennarelli (1974) could produce cortical contusions by translation and rotation but they were more discrete and superficial in the translated group. Widespread superficial and deep lesions on the other hand were only seen in the rotated animals.

The influence of head impact was explored in the rhesus monkey by Hirsch *et al.* (1970). With direct occipital impacts the levels of acceleration required to produce concussion were only half of those required when only whiplash trauma was used.

The hypothesis that widespread shear strains, elicited by angular motion, would be the critical factor for concussion thus seemed to be supported. Ommaya and Gennarelli (1974) also emphasized that concussion includes "a graded set of clinical syndromes following head injury wherein increasing severity of disturbance in level and content of consciousness is caused by mechanically induced strains affecting the brain in a centripetal sequence of disruptive effects of function and structure". This, the *centripetal hypothesis of cerebral concussion* is illustrated in Fig. 7. The transient dysfunctions are defined by a continuous spectrum of specific disturbances: confusion without amnesia (stunning), confusion and amnesia of slow or rapid onset, coma of varying length, and, concomitant with the states of unconsciousness, sensory motor paralyses, disturbances in respiration and circulation (Ommaya 1985). As a support for this hypothesis neuropathological as well as clinical data have been advanced.

The clinical observations that the "return of awareness to stimuli usually precedes motor and sensory recovery which in turn recover before restoration of memory and other cognitive functions" (Ommaya and Gennarelli (1974) supports that the cortex, limbic and frontal areas in particular, are more vulnerable than deeper parts of the brain. Moreover the amnesic component of the concussion syndrome is a more sensitive indicator than the paralytic phenomena. Confusion and amnesia can occur without a patient being unconscious but the contrary does not apply (Yarnell and Lynch 1973).

D.A. Stålhammar

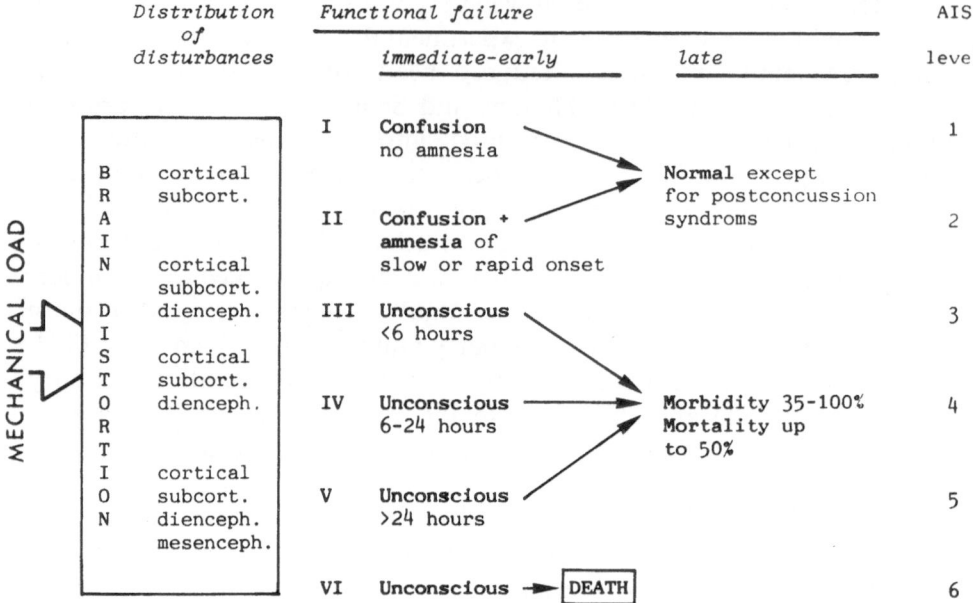

Fig. 7. Hypotheses for the syndromes of cerebral concussion. Mechanical load produces, in varying distribution, brain distorsion which cause different structural disturbances proceeding from the cortex inwards to the mesencephalic core. Functional rather than structural failure in this order is implied. Tissue damage occurs in a related but not identical distribution. Grade of concussive brain injury is correlated with injury severity according to the Abbreviated Injury Scale (*AIS*) 0–1 no or minor, *AIS* 2 moderate, *AIS* 3 serious, *AIS* 4 severe, *AIS* 5 critical and *AIS* 6 fatal injury. (Modified from Ommaya and Gennarelli 1974 and Ommaya 1985 and reprinted from Stålhammar[53] by permission)

The non-occurrence of primary brain stem damage without widespread lesions in cortical and subcortical areas in animal experiments (Ommaya and Gennarelli 1974) is paralleled by similar findings in human autopsy material (Mitchell and Adams 1973, Adams *et al.* 1977).

Ommaya and Gennarelli (1974) also used somatosensory evoked responses to investigate the disturbances generated by angular and translational motion. Specific disturbances occurred in the animals subjected to angular acceleration indicating that subcortical and cortical functions were always severely disturbed during paralytic concussion, *i.e.* when the deformation had reached and eliminated the function of the "rostral part of the diencephalic-mesencephalic core".

Later it was evidenced that the pathology found in humans after severe head injuries closely resembled the structural alterations seen in animal experiments (Adams *et al.* 1982a). Gennarelli *et al.* (1982a) studied the effects of varied, distributed load with angular acceleration and demonstrated that transient and prolonged traumatic unconsciousness, similar to that in humans, could be produced. The length and the depth of the coma were related to the magnitude, duration, and direction of the acceleration. Laterally directed angular acceleration generated longer and deeper coma, as well as widespread axonal damage, focal lesions in the corpus callosum and in the area of the superior cerebellar peduncle, so-called "diffuse axonal injuries" (DAI, Gennarelli *et al.* 1982a) than comparable levels of sagittal acceleration.

Pressure Gradient – Translational Acceleration Theory

The damaging effects of translational acceleration have been attributed to the induction of intracranial pressure differences, which may be generated by the propagation of stress waves from the site of impact and may result in motion of the intracranial contents, absolute and relative to the skull. These motions may produce damage at pressure differences in directions not parallel to the impact direction, *e.g.* towards the anterior skull base, foramen magnum and the spinal canal. Lindgren (1966) recorded pressure differences over the cranio-spinal junction in human cadaver skulls subjected to impacts and therefore hypothesized that brain stem "flow" could be generated. Although some other investigators also favored this idea, 'shear stress' in the brain stem caused by pressure gradients for the production of functional brain stem disturbances has been questioned.

Cavitation

In areas with negative pressure transients injurious strains may be produced by so-called cavitation (Lindgren 1966, Sellier and Unterharnsheidt 1963, Goldsmith 1966, 1983) although it was doubted whether such very short negative pressure pulses actually would generate cavitation bubbles in the viscoelastic brain tissue and blood (von Gierke 1966 and Goldsmith 1972).

In a series of animal experiments the present author tested the cavitation hypotheses (Stålhammar 1975, Stålhammar and Olsson 1975a,b). However, no injuries at all were found in brain areas subjected only to negative pressure pulses of about − 1 atm and thus it was concluded that some other factors might be added to explain the occurrence of cortical contusions (*cf.* Tsubokawa *et al.* 1975). These findings would be in accordance with Lubock and Goldsmith (1980) who produced cavitation bubbles, at coup and countre-coup sites, in various kinds of liquids (water, glycerin, artificial CSF) but not in gelatin, a colloidal suspension similar to neural tissue, and Goldsmith (1983) concluded that brain damage might occur from cavitation. Also other authors claim that translational acceleration does contribute to subdural hematoma, contusions remote from site of impact and to some of the focal contusions (Ommaya 1985, Gennarelli 1985).

Cranio-spinal Junction Movements

Flexion–extension and/or *bending in the cranio-spinal junction* has been claimed as a separate mechanism explaining cerebral concussion. Friede 1961 and von Gierke 1966 demonstrated stretching of the cervical cord around the odontoid process with concomitant focal pathomorphological changes and concussion. However, these effects probably result from shear stresses transmitted from the spinal cord up into the brain stem.

Primary Injuries

Primary injuries mean immediate impact injuries, such as fractures, contusions, lacerations and widespread damage to the white matter, and primary complications; epi- and subdural bleedings and brain swelling (Adams *et al.* 1980). These injuries can result from the effects of either mainly contact or of mainly inertial forces (Table 1). According to their distribution these can be categorized as *focal brain injuries* (contusions and hematomas) or *diffuse brain injuries* (axonal injury, brain swelling and petechial hemorrhages). Several primary injuries occur simultaneously, although one particular type usually dominates.

Fractures

Fractures, epidural hematomas and some brain contusions are entirely caused by the skull deformation produced by the contact forces *per se*, at site of impact or at other locations. The biomechanics of skull distortion and fracturing has been studied by Gurdjian and collaborators (Gurdjian *et al.* 1947) and by Melvin and Evans 1971. At blunt impact the inbending at the point of loading causes an outbending in the surrounding area, which exerts a tensile stress on the outer table of the calvarium eventually leading to a failure. The fractures, occurring particularly in thinner areas, spread along areas of the least resistance. In the skull base transverse fractures can be produced from side to side, and antero-posterior fractures in frontal impacts may be accompanied by injury to the pituitary or to the optic chiasm.

Table 1. *Mechanisms of Primary Injuries.* Focal and diffuse primary brain injuries, occurring immediately at the impact or as a primary complication, and their relation to the contact or the inertial forces of the mechanical load. (Reprinted from Stålhammar[53] by permission)

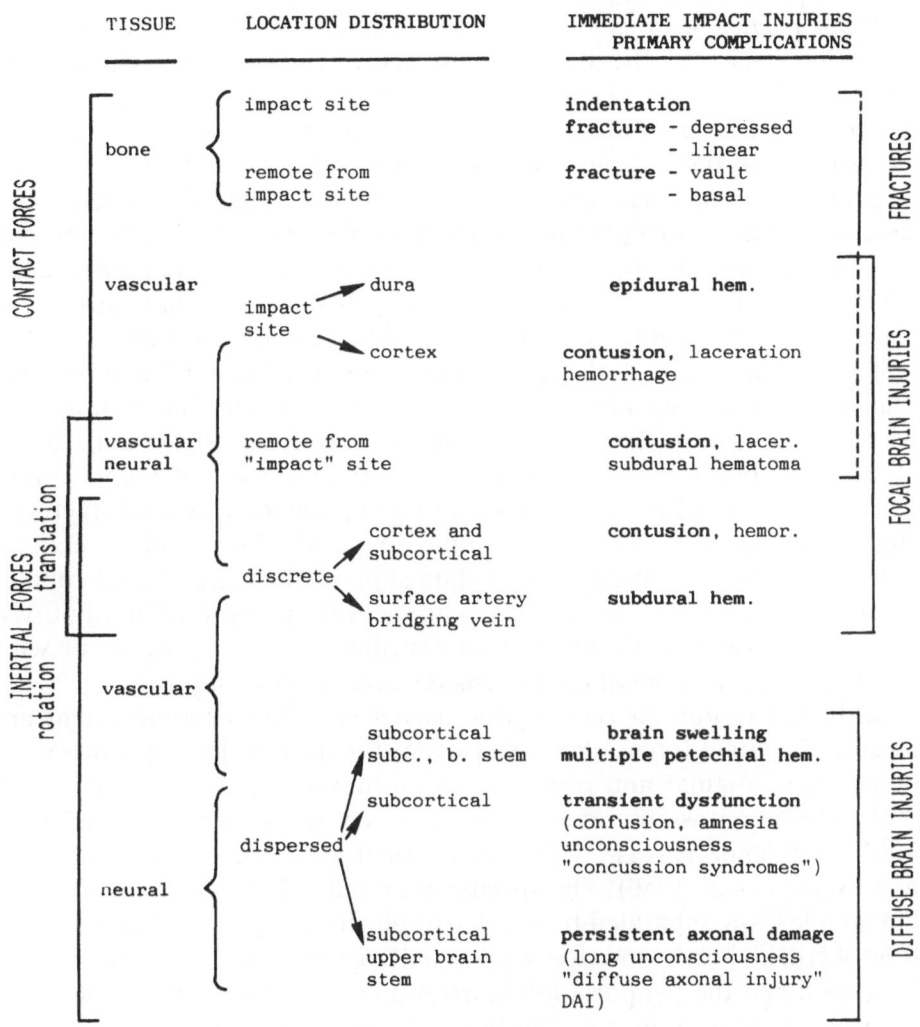

Focal Injuries

Epidural hematomas (EDH) are complications related directly to a deformation of the skull bone at which the dural fibrous and vascular attachment to the inner table is sheared. Thereby an extradural cavity can be created from a bleeding artery or vein or more diffusely from the bone and dural surfaces. Although EDH statistically is most often associated with skull fractures, over all ages in 80–90% (Guillermain 1986), the consequences of fractures *per se*, by tearing a dural vessel, do not seem to be the principal cause (Hooper 1959).

Brain contusion at site of impact is caused by the inbending bone generating strains in the vascular and neural tissue as demonstrated in many experiments (Tornheim *et al.* 1983). The injuries beneath the area of loading have also been attributed to the tensile strain ('negative pressure') occurring when the indented bone snaps back (Lindgren 1966).

Brain contusion remote from impact site can be related to bone distorsion generated by the contact forces and to the strains elicited by inertial forces. Such injuries have been described directly beneath areas of bone where the deformation has caused fractures. Sjöwall (1943) found contusions in the inferior part of the frontal lobe adjacent to fractures in the orbital roof after a blow to the back of the head and along meridional fractures also discussed by Lindgren 1966. Tornheim *et al.* (1983), after blunt trauma to the head of the cat, also demonstrated contusions contralateral to the impact but always with concomitant fractures. In clinical studies more severe contusions have been demonstrated in patients with skull fractures, while fewer fractures have been found in patients with diffuse axonal injuries (Adams *et al.* 1982a). These data support the concept that contact forces do contribute to contusions, while severe and fatal diffuse injuries, caused by angular acceleration from distributed impact may occur with only occasional and small contusions (Gennarelli *et al.* 1982a).

Although *contusions remote from impact* certainly to some extent are a result of contact forces, there is evidence for their major dependence on the effects of angular and translational motion (Ommaya and Gennarelli 1974). These contusions were always more severe in the frontal and temporal lobe (Gennarelli *et al.* 1979) and considered identical to those found in man (Adams *et al.* 1980). The specific anatomical features of the anterior and middle fossa, separated by the sharp sphenoid ridge, have been claimed to be of critical importance for the dominating appearance of contusions in the frontal and the temporal lobes, irrespective of whether the patient has received a frontal or an occipital blow (Gurdjian *et al.* 1955, Adams *et al.* 1980).

Small *deep hematomas* in the cerebral hemispheres are probably attributable to enhanced levels of localized strain produced by inertial forces mainly from rotation.

Contusions parasagittally at the upper margin of the brain (so-called gliding contusion) has been described in man and in experiments on animals subjected to sagittal angular acceleration (Adams *et al.* 1980, Löwenhielm 1975).

Subdural hematoma (SDH) associated with brain contusions, and produced by the combined effects of contact phenomena and inertia, comprise about half of these bleedings, the rest being 'pure' SDH, resulting exclusively from brain movements tearing surface vessels *i.e.* arteries and bridging veins (Teasdale and Galbraith 1981).

SDH caused by injured surface arteries are reported in about 10% of all SDH and are most often associated with less severe trauma (Drake 1961, Shenkin 1982). In translated squirrel monkeys some focal subdural hematoma were produced while in the rotated group much more pronounced subdural hematoma were found in all animals (Ommaya and Gennarelli 1974). Subdural hematomas were also pronounced in the sagittally rotated monkeys proportionate to the rotational acceleration although contact forces clearly contributed in these experiments, since 14 of 49 animals in groups with SDH presented with fractures (Adams *et al.* 1982b).

Diffuse Injuries

Subdural hematoma and diffuse injuries. The parallels between subdural hematomas on the one hand and the severity of concussion syndrome and widespread axonal damage on the other are not always seen. Gennarelli and Thibault (1982) subjected the rhesus monkey to a sudden, distributed, sagittal plane angular motion composed of an acceleration–deceleration pulse, the latter being more pronounced. By increasing the rise time of the deceleration phase, lowering its peak magnitude and prolonging its duration, prolonged coma and widespread axonal injury were produced without concomitant acute subdural hematoma (Fig. 8). If the pulse duration and the rise time were shortened at about the same level of peak angular deceleration the animals had a subdural hematoma and most of them were also concussed.

The explanation was advanced that the bridging veins are particularly sensitive to a high strain rate which does not generate deep tissue strains severe enough to produce long coma and severe axonal damage. These findings would be in accordance with clinical experience where SDH are significantly more common from a fall and assault than at motor vehicle accidents (Gennarelli 1984); strain rates in the former types of accident are probably higher than in the latter, where the head is subjected to a load of longer duration (Gennarelli and Thibault 1982). However, an analysis of these experiments, by Lee *et al.* (1987), emphasized that the acceleration phase of the load as well as the translational component of the motion have to be considered regarding the biological response.

The etiology of *multiple petechial hemorrhages* in the white matter of the frontal and temporal lobes and particularly in the rostral brain stem, seen in

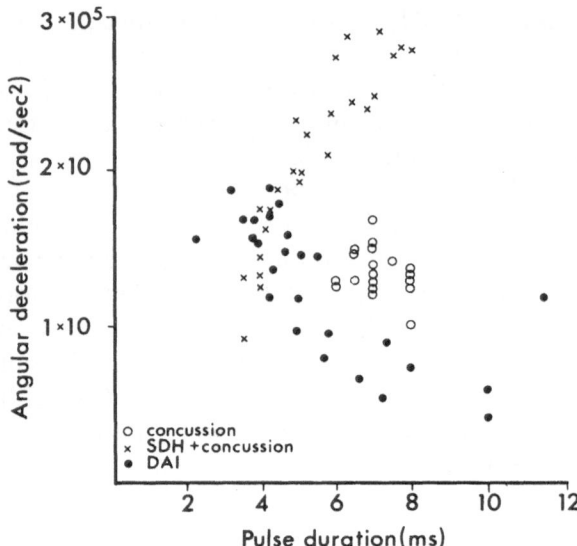

Fig. 8. Concussion, diffuse axonal injury (DAI) and subdural hematoma (SDH) with concomitant concussion related to the peak and the duration of the deceleration phase at angular motion in the rhesus monkey. (Modified from Gennarelli and Thibault 1982 and reprinted from Stålhammar[53] by permission)

patients dying very soon after violent accidents (Tomlinson 1970), is unclear. Ommaya and Gennarelli (1974) found more scattered petechial hemorrhages in the rotated group.

Diffuse *brain swelling* is well known from clinical experience; in the surroundings of contusions and hematomas, and in one or both hemispheres. Unilateral swelling is particularly associated with acute subdural hematoma in the human (Adams *et al.* 1980), and also in experimental animals subjected to non-deforming angular acceleration (Adams *et al.* 1982b). Bilateral diffuse swelling has been described more often in children and adolescents. Vasomotor paralysis, is probably caused by the widespread changes in the brain producing immediate and lasting coma. However, so far the mechanical events responsible for these injuries have not been clarified.

The pathogenesis of the *concussion syndromes* including *widespread damage to axons and vasculature* is discussed above (p 11). The persistent structural damage of this kind has been denoted "shearing injuries" (Strich 1956) and "diffuse axonal injuries", "DAI" (Adams *et al.* 1982a). Similar injuries have also been produced in primates where it is related to the magnitude of the rotational acceleration, more easily generated by oblique and lateral than sagittal motions (Gennarelli *et al.* 1982a).

Aspects of Injury Reduction

The application of biomechanics to define brain tolerance levels for the purpose of injury mitigation will be briefly described.

The magnitude of functional and/or structural changes as a result of mechanical violence is denoted by 'injury severity level'. The Glasgow Coma Scale sum score (Teasdale and Jennett 1974), the Reaction Level Scale (RLS85) (Starmark *et al.* 1988), type of intracranial lesion (Gennarelli *et al.* 1982b) and the Abbreviated Injury Scale (AIS) (AIS 1985) are measures commonly used.

The injury severity level must be correlated with a physical parameter, such as rotational and linear acceleration. The magnitude of mechanical loading (expressed by acceleration) that produces a specific type and severity of injury is called the *'tolerance level'*.

Since severe dysfunction of the brain can occur at much lower levels of loading than at those resulting in structural damage, it is problematic to establish valid and practical tolerance levels for brain failures. Indeed, separate tolerance levels for different areas and types (*cf.* Fig. 2) may be required.

The use of *translational acceleration*, as a head injury criterion, originates from experimental studies in human cadavers in the relation between acceleration (pulse duration 1–6 ms), intracranial pressure and the production of linear skull fractures (Lissner *et al.* 1960). The curve displaying these relationships was later extended to pulse durations above 6 ms by comparison with animal and human volunteer experiments and known as the Wayne State Tolerance Curve (WSTC), aimed to delineate the level of concussion (Fig. 9).

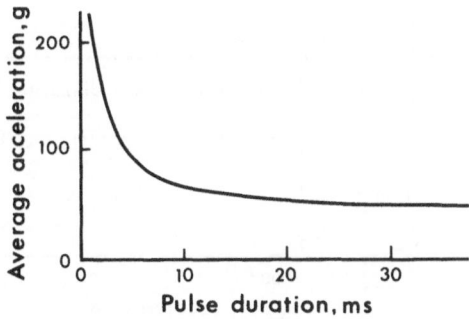

Fig. 9. Relations between acceleration and time required to produce a fracture at frontal skull impact in the human cadaver according to Lissner *et al.* 1960 and revised to the Wayne State Tolerance Curve delineating cerebral concussion. (Modified from Goldsmith 1983 and reprinted from Stålhammar[53] by permission)

The Wayne State Tolerance Curve was found to be an approximately straight line in a log-log plot, and thus the Severity Index (SI) (SI = I $a^{2.5}$dt) was created (Gadd 1967). SI $\leqslant$ 1.000 was considered acceptable. Later the acceleration term in the expression for SI was replaced by a time averaged and weighted acceleration since SI was considered too conservative for non-contact whiplash loading. By this improvement, the rate of load application was included and the Head Injury Criterion (HIC) was derived. The value HIC = 1.000, was established in 1972 in the US, as a legal limit of great practical importance.

$$HIC = \left(\frac{1}{t_2 - t_1} \int_{t_1}^{t_2} a\,dt \right)^{2.5} (t_2 - t_1)$$

HIC can be understood as a measure of the rate of change of specific kinetic energy (V^2/T) modulated by the square root of the average acceleration over the time interval T ($\sqrt{V/T}$).

HIC cannot reliably be applied for brain injury caused by rotational acceleration and by impacts generating contact forces above the limits for skull fracture. Moreover, a continuous criterion would be more conveniently applied than HIC which is an 'either/or' limit.

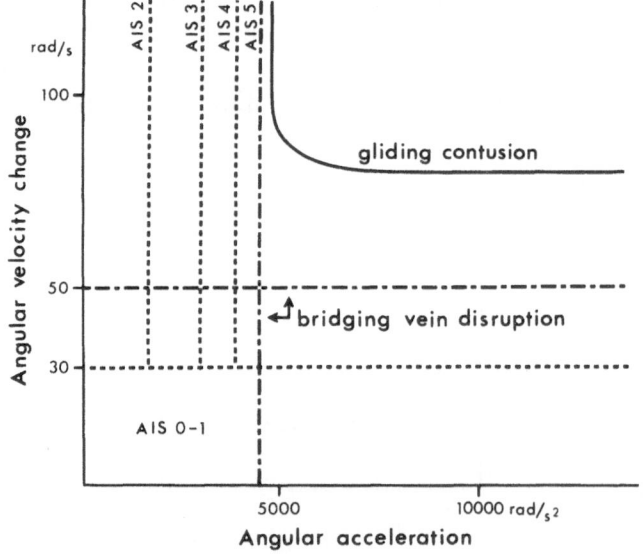

Fig. 10. Proposed tolerance limits: for gliding contusions (Löwenhielm 1975), bridging vein disruption (Löwenhielm 1974b), safe area: to the left and under these limits, and for various degrees of injury severity according to AIS (Goldsmith and Ommaya 1984). (Reprinted from Stålhammar[53] by permission)

The *Mean Strain Criterion* (MSC) is such a continuous criterion suggested specifically to account for the contact and translation components of the impact (Stalnaker *et al.* 1987). The MSC is based directly on experiments on a physical and theoretical head injury model which has been validated on living primates and cadavers from AIS 0 to 6, *i.e.* from no injury to fatal injury.

Since MSC is most valid for direct head impacts and inertial loading by translational acceleration, it should be complemented by *injury criteria describing rotational motion*, since such load is required to produce transient and persistent unconsciousness, diffuse axonal injuries and rupture of bridging veins (*cf.* Table 1). Injury criteria of this kind are illustrated in Fig. 10, and as seen, the duration of loading decides which will be the most appropriate parameter.

References

1. Abbreviated Injury Scale 1985 Revision. American Association for Automotive Medicine, Arlington Heights IL. 60005. USA
2. Adams JH, Mitchell DE, Graham DI, Doyle D (1977) Diffuse brain damage of immediate impact type. Its relationship to 'Primary brain-stem damage' in head injury. Brain 100: 489–502
3. Adams JH, Graham D, Scott G, Parker L, Doyle D (1980) Brain damage in fatal non-missile head injury. J Clin Pathol 33: 1132–1145
4. Adams JH, Graham DI, Murray LS, Scott G (1982a) Diffuse axonal injury due to nonmissile head injury in humans: An analysis of 45 cases. Ann Neurol 12: 557–563
5. Adams JH, Graham DI, Gennarelli TA (1982b) Neuropathology of acceleration-induced head injury in the subhuman primate. In: Grossman RG, Gildenberg PL (eds) Head injury: basic and clinical aspects. Raven Press, New York, pp 141–150
6. Aldman B, Ljung C, Thorngren L (1982) Intracranial deformation patterns due to impulsive loading – A model study. Paper for presentation at technical Session No 2 Biomechanics and Dummy Development of the 9th ESV Conference in Kyoto, November 3, 1982
7. Drake CG (1961) Subdural haematoma from arterial rupture. J Neurosurg 18: 597–601
8. Friede RL (1961) Experimental acceleration concussions. Arch Neurol (Chicago) 4: 449–462
9. Gadd CW (1967) Use of a weighted impulse criterion for estimating injury hazard. In: 10th Stapp Car Crash Conference Proceedings, 8–9 November 1966, Alamogordo NM, New York: SAE paper no. 660793, pp 164–174
10. Gennarelli TA (1984) Clinical and experimental head injury. In: Aldman B, Chapon A (eds) The biomechanics of impact trauma. Elsevier Science Publishing Company, Amsterdam, pp 103–115

11. Gennarelli TA (1985) The state of the art of head injury biomechanics. A review. Proceedings from the 29th Annual Conference, Washington, DC, October 7–9, 1985

12. Gennarelli TA, Thibault LE (1982) Biomechanics of acute subdural hematoma. J Trauma 22: 680–686

13. Gennarelli TA, Abel JM, Adams H, Graham D (1979) Differential tolerance of frontal and temporal lobes to contusion induced by angular acceleration. Proceedings of Twenty-third Stapp Car Crash Conference, October 17–19, San Diego, California. Published by Society of Automotive Engineers, Inc. 400 Commonwealth Drive, Warrendale, Pennsylvania 15096, pp 563–586

14. Gennarelli TA, Thibault LE, Adams JH, Graham DI, Thompson CJ, Marcincin RP (1982a) Diffuse axonal injury and traumatic coma in the primate. Ann Neurol 12: 564–574

15. Gennarelli TA, Spielman GM, Langfitt TW, Gildenberg PL, Harrington T, Jane JA, Marshall LF, Miller JD, Pitts LH (1982b) Influence of the type of intracranial lesion on outcome from severe head injury. J Neurosurg 56: 26–32

16. von Gierke HE (1966) On the dynamics of some head injury mechanisms. In: Caveness WF, Walker AE (eds) Head injury. Conference proceedings. JB Lippincott Company, Philadelphia Toronto, pp 383–396

17. Goldsmith W (1966) The physical processes producing head injuries. In: Caveness WF, Walker AE (eds) Head injury. Conference proceedings. JB Lippincott Company, Philadelphia Toronto, pp 350–382

18. Goldsmith W (1972) Biomechanics of head injury. In: Fung YC, Perrone N, Anliker M (eds) Biomechanics. Its foundation and objectives. Prentice-Hall, Englewood Cliffs, NJ, pp 585–634

19. Goldsmith W (1983) Some aspects of the physical and mathematical modeling of loading to head/neck systems and implications of current DOT injury criteria. In: Head and neck injury. A Consensus Workshop, Department of Transportation HS 806 434. Government Printing Office, Washington, DC pp 133–148

20. Goldsmith W, Ommaya AK (1984) Head and neck injury criteria and tolerance levels. In: Aldman B, Chapon A (eds) The biomechanics of impact trauma. Elsevier Science Publishers B.V. Amsterdam, pp 149–187

21. Guillermain P (1986) Traumatic extradural hematomas. In: McLaurin RL (ed) Advances in neurotraumatology. Extracerebral collections. Springer, Wien New York, pp 1–50

22. Gurdjian ES, Lissner HR, Webster JE (1947) The Mechanism of production of linear skull fracture. Further studies on deformation of the skull by the "stresscoat" technique. J Surg Gynecol Obstet 85: 195–210

23. Gurdjian ES, Webster JE, Lissner HR (1955) Observations on the mechanism of brain concussion, contusion, and laceration. J Surg Gynecol Obstet 101: 680–690

24. Higgins LS, Schmall RA (1967) A device for the investigation of head injury effected by non-deforming head acceleration. Proc 11th Stapp Car Crash Conference, Society of Automotive Engineers, Inc., New York, pp 57–72

25. Hirsch A, Ommaya A, Mahone R (1970) Tolerance of subhuman primate brain to cerebral concussion. Chapter XVI. In: Gurdjian ES, Lange WA, Patrick LM, Thomas LM (eds) Impact injury and crash protection. Ch C Thomas, Springfield, Illinois, pp 352–369

26. Hodgson VR (1967) Tolerance of the facial bones to impact. Am J Anat 120: 113–122

27. Holbourn AHS (1943) Mechanics of head injuries. Lancet 2: 438–441

28. Hooper R (1959) Observations on extradural haemorrhage. Br J Surg 47: 71–87

29. Kornhauser M (1954) Prediction and evaluation of sensitivity to transient accelerations. J Appl Mech 21: 371–380

30. Kornhauser M, Gold A (1962) Application of the impact sensitivity method to animate structures. Proceedings of a Symposium: Impact Acceleration Stress. Washington, National Academy of Sciences, National Research Council, pp 333–344

31. Lee MC, Melvin JW, Ueno K (1987) Finite element analysis of traumatic subdural hematoma. 31st Stapp Car Crash, Society of Automotive Engineers, Inc. 400 Commonwealth Drive Warrendale, PA 15096–0001, Conference Proceedings, pp 67–77

32. Lindgren SO (1966) Experimental studies of mechanical effects in head injury. Acta Chir Scand [Suppl] 360: 1–100

33. Lindgren S (1983) Mechanical and clinical approaches to head injury criteria. In: Head and neck injury Criteria. A Consensus Workshop, Department of Transportation HS 806 434. Government Printing Office, Washington, DC, pp 49–61

34. Lissner HR, Lebow M, Evans FG (1960) Experimental studies on the relation between acceleration and intracranial pressure changes in man. Surg Gynec Obstet 111: 329–338

35. Liu YK (1979) Biomechanics and biophysics of CNS trauma. In: Odom GL (ed) Central nervous system trauma research. Status report, pp 36–52

36. Ljung C, Lindgren S, Aldman B (1983) On the analytical approach to head injury criteria. Proceedings of a Workshop on Head and Neck Injury Criteria. US Department of Transportation. National Highway Traffic Safety Administration. Washington, DC, pp 194–197

37. Löfgren J, Pellettieri L (1967) Epidemiologia della contusione del lobo temporale. Minerva Neurochir 11: 242–246

38. Lubock P, Goldsmith W (1980) Experimental cavitation studies in a model head-neck system. J Biomech 13: 1041–1052

39. Löwenhielm P (1975) Mathematical stimulation of gliding contusions. J Biomech 8: 351–356

40. Melvin JW, Evans FG (1971) A strain energy approach to the mechanics of skull fracture. Proc 15th Stapp Car Crash Conference, Society of Automotive Engineers, Inc, New York, pp 666–685

41. Mitchell DE, Adams JH (1973) Primary focal impact damage to the brainstem in blunt head injuries. Does it exist? Lancet 4: 215–218

42. Ommaya AK (1968) Mechanical properties of tissues of the nervous system. J Biomechanics 1: 127–138

43. Ommaya AK (1985) Biomechanics of head injury: Experimental aspects. In: Nahum AM, Melvin J (eds) The Biomechanics of Trauma. Appleton-Century-Crofts, Norwalk, Connecticut, pp 245–269

44. Ommaya AK, Gennarelli TA (1974) Cerebral concussion and traumatic unconsciousness: Correlation of experimental and clinical observations on blunt head injuries. Brain 97: 633–654

45. Pudenz RH, Shelden HC (1946) The lucite calvarium. A method for direct observation of the brain. II. Cranial trauma brain movement. J Neurosurg 3: 487–505

46. Sellier K, Unterharnscheidt F (1963) Mechanik und Pathomorphologie der Hirnschäden nach stumpfer Gewalteinwirkung auf den Schädel. Springer, Berlin Heidelberg New York

47. Shenkin HA (1982) Acute subdural hematoma. Review of 39 consecutive cases with high incidence of cortical artery rupture. J Neurosurg 57: 254–257

48. Sjövall H (1943) The genesis of skull and brain injuries. An anatomical and physical study. Munksgaard, Copenhagen

49. Stalnaker RL, Low TC, Lin AC (1987) Translational energy criteria and its correlation with head injury in the sub-human primate. Proceedings of the 1987 International Conference on the Biomechanics of Impact, Birmingham, pp 223–238

50. Strich SJ (1956) Diffuse degeneration of the cerebral white matter in severe dementia following head injury. J Neurol Neurosurg Psychiatry 19: 163–185

51. Starmark JE, Stålhammar D, Holmgren E (1988) The Reaction Level Scale (RLS85). Manual and guidelines. Acta Neurochir (Wien) 991: 12–20

52. Stålhammar D (1975) Experimental brain damage from fluid pressures due to impact acceleration. 1. Design of experimental procedure. Acta Neurol Scand 52: 7–26

53. Stålhammar D (1990) Mechanism of brain injury. In: Winken PJ, Bruyn GW, Klawans HL (eds) Head injury. Handbook of clinical neurology. Elsevier Science Publishers B. V., Amsterdam, in press

54. Stålhammar D, Olsson Y (1975a) Experimental brain damage from fluid pressures due to impact acceleration. 3. Morphological observations. Acta Neurol Scand 52: 38–55

55. Stålhammar D, Olsson Y (1975b) Experimental brain damage from fluid pressures due to impact acceleration. 4. Comparative studies with acceleration–concussion. Acta Neurol Scand 52: 94–110

56. Teasdale G, Galbraith S (1981) Acute traumatic intracranial hematomas. Progress in Neurological Surgery 10: 66–99

57. Teasdale G, Jennett B (1974) Assessment of coma and impairment of consciousness. A practical scale. Lancet ii 81–84

58. Thibault LE, Gennarelli TA (1985) Biomechanics and craniocerebral trauma. In: Becker DP, Povlishick JT (eds) Central nervous system trauma. Status report, National Institute of Neurological and Communicative Disorders and Stroke, National Institutes of Health, pp 379–389

59. Tomlinson BE (1970) Brain-stem lesions after head injury. J Clin Pathol 23 [Suppl 4]: 154–165
60. Tornheim PA, Liwnicz BH, Hirsch CS, Brown DL, McLaurin RL (1983) Acute responses to blunt head trauma. Experimental model and gross pathology. J Neurosurg 59: 431–438
61. Tsubokawa T, Nakamura S, Hayashi N, Miyagami M, Taguma N, Yamada J, Kurisaka M, Sugawara T, Shinozaki H, Goto T, Takeuchi T, Moriyasu N (1975) Experimental primary fatal head injury caused by linear acceleration – biomechanics and pathogenesis. Neurol Med Chir 15: 57–65
62. Unterharnscheidt F, Higgins LS (1969) Traumatic lesions of brain and spinal cord due to nondeforming angular acceleration of the head. Texas Reports on Biology and Medicine 27: 127–166
63. Yarnell PR, Lynch S (1973) The 'ding': Amnestic states in football trauma. Neurology 23: 196–197

Cerebral Hemispheric Contusions and Lacerations

I. Oprescu*

Neurosurgical Clinic, Hospital G. Marinescu, Bucharest (Romania)

With 28 Figures

Contents

* Most regretfully Ion Oprescu died on the 12th December 1988. Therefore his manuscript and figures are nearly the same as his original version. For revision of terminology we thank Phillip Harris, Edinburgh, and Roland Schröder, Köln.

Introduction

The limits, difficulties and some controversies in our understanding of the neuropathology of the injured brain have been competently discussed by a number of authors, and more recently by Strich (1969, 1970), Adams and Graham (1972), Adams (1975), Adams *et al.* (1977), Vigouroux and Guillermain (1981), Jennett and Teasdale (1981), Arseni and Oprescu (1972, 1984), Henn (1989).

The material for study is usually obtained by autopsy rather than by biopsy performed during surgery, compared to patients with brain lacerations and for some with hematomas. The autopsy material presents certain drawbacks. One of the more important may be attributed to the medico-legal conditions as the autopsy in all head injuries must be performed by forensic pathologists, and is therefore not carried out sooner than 6 to 12 hours and in some cases even 24 hours after death. Thus structural alterations may be superimposed onto the traumatic lesions, rendering a minute study of the latter very difficult. The fact that the public prosecutor is obliged to draw up a statement on the causes of death within a relatively short interval, usually obliges the neuropathologist to evaluate the main lesions on the fresh, non-fixed brain. To overcome this disadvantage we were able to obtain, following an agreement with the Medico-Legal-Institute of Bucharest, permission to study together the lesions in a series of 200 cases, after correct fixation of the injured brains.

Another important difficulty was brought about by modern intensive care techniques, which are able to maintain patients with brain injuries alive for a rather long interval, theoretically incompatible with survival. This has led to the listing of some new neurotraumatological entities, for instance prolonged vegetative state and other types of posttraumatic encephalopathies; but also the reverse: maintenance of the brain under a precarious survival condition which is apt to induce superadded lesions, hence vitiating the accuracy of the study.

A third difficulty seems to be the fact that the material from autopsy actually represents the final stage of a traumatic brain lesion with a fundamental evolutive character, and which therefore passed through a number of phases succeeding one another at a slower or more rapid rate. These intermediate stages, which are of particular interest, inevitably escape the research worker who needs to reconstitute their sequences.

A possible method to reconstitute these stages, attempted in our studies, was the use of material obtained from cases of multiple injuries with minor, mean or severe brain damage, but in which death was caused by predominantly extracerebral, somatovisceral lesions. In these cases too, the study can only reveal a single moment in the evolution of a brain lesion.

A second attempt at correcting the vicissitudes inherent to autopsy material was the series of experimental studies carried out especially in the last four decades, given in the references in the report by Ommaya and Gennarelli (1975). The relative value of the experimental models and their critical evaluation for clinical purposes, is to be found in the work of Jennett and Teasdale (1981).

The place of cerebral contusions and lacerations in the framework of head injuries.

It must be emphasized that neither contusions nor lacerations may exist as a single, isolated traumatic effect.

Brain contusions, irrespective of their severity, are always accompanied by a degree of cerebral edema, and also often by intracranial blood or fluid effusions as hematomas and/or the so-called meningitis serosa.

Brain lacerations result from a major blow to the head and in most cases there is complex pathophysiology with perifocal contusion and edema or extracerebral blood effusions.

Although brain contusions and lacerations have an evolutive character, there is usually a dominant lesion and this must be considered. Such a classification is necessary for discussions, for the elaboration of a correct diagnosis, and also as a valuable adjuvant for establishing therapeutic priorities and prognosis. Finally, it offers the advantage of comparing the results obtained from different neurosurgical units.

In order to determine the place of contusions and lacerations amongst the cerebral traumatic effects, we use a personal classification, (Table 1) considering especially the lesions of the brain itself, and, to a lesser extent, the physiopathological and clinical criteria (Arseni and Oprescu 1977).

In this classification, contusions and lacerations are considered as immediate *primary* non-mediated and specific posttraumatic effects. In contrast, the *secondary* effects (intracranial blood or fluid effusions) are not strictly specific, but can appear as a non-obligatory consequence of the primary effects. Also in contrast, the *subsequent* effects such as cerebral edema and cerebro-ventricular collapse are not specific to the trauma, but rather appear as accompanying lesions.

Table 1. *A Classification of the Posttraumatic Cerebral Effects*

Immediate and recent

Primary	Secondary	Subsequent
Cerebral concussion	Blood effusions	Brain oedema
Cerebral contusions	– epidural	Brain collapse
– minor	– subarachnoid	
– medium	– subdural	
– severe { hemispheric / Brain stem	– intracerebral	
	Fluid effusions	
	– Meningitis serosa	
Cerebral lacerations		
– direct		
– indirect		

Late posttraumatic cerebral effects

Evolutive:	Posttraumatic encephalopathies
	Meningo-cerebral scars
Sequelae:	Hemiplegia, Aphasia etc.

Table 2. *Summary of Neuropathological Findings and Pathophysiology of Post-traumatic Cerebral Contusions and Lacerations*

Cerebral Hemispheric Contusions and Lacerations

Investigations, Definitions/   **Neuropathological Findings and Pathophysiology**

A. Contusions

1. Immediate cerebral effects		30
Macroscopic pathology		31
Diffuse contusions	both hemispheres involved	
Limited contusions	one lobe involved	
Focal contusions	preferential location: fronto-orbito-temporal	
	exceptional location: hypothalamic	
	special location: contre-coup	
	heterotopic location	
Microscopic pathology		36
Vascular alterations	vasodilatation, hemorrhage, thrombosis	37
Parenchymal lesions	neuronal: following ischemia, ICP elevation, degeneration in deep structures and cerebellum	
	microglial: scaverenger cells, microglial stars	41

Table 2 (continued)

	astroglial:	hypertrophy
	oligodendroglial:	diminished density
	fibers:	Wallerian degeneration 42
2. Late cerebral effects		43
Encephalopathies	mixed with pre- and coexisting non-traumatic factors, as arteriosclerosis, chronic hypoxia	43
Evolutive type	hydrocephalus protracted form (Jellinger) punch-drunk encephalopathy	45
Non-evolutive type	sequelae	
Macroscopic pathology	hemispheric asymmetry asymmetric hydrocephalus old infarctions thinning of corpus callosum infection areas	45
Microscopic pathology	diffuse demyelination sparing associate fibers by traumatic axonal interruption, hypoxia, edema cortical, thalamic and cerebellar neuronal rarefaction microscars neurofibrillary tangles of neurons	45

B. Lacerations

1. Immediate effects

Types: direct	penetrating wounds	46
indirect	shearing of tissue of different inertia	
secondary	by hematoma	
Macroscopic pathology	parenchymatous disruption	77
Microscopic pathology	necrosis	49
	blood effusion microextravasations	

2. Late effects

Meningo-cerebral scar		54
Macroscopic pathology	meningeal scar	54
	meningo-cortical scar	55
	meningo-cerebro-ventricular scar	
Microscopic pathology	central collagen-riche connective tissue	55
	intermediate mesenchymal-glial plait	
	peripheral areas of reactive glia and demyelination	
Encephalopathy	as in A2	

The *late* posttraumatic effects are the traumatic encephalopathies, and the meningo-cerebral scars, usually a consequence of brain lacerations. Both have a slow evolutive character (Table 2).

Contusions

Immediate Cerebral Effects

In autopsy material contusions are the most frequent traumatic cerebral lesions and may occur in approximately 80% of all closed head injuries, the remaining 20% including concussions and lacerations as other primary lesions and the secondary and subsequent ones.

Characteristic of the immediate primary effect of a contusion is the marked polymorphism which depends upon the qualities of the impact agent and the biomechanic and pathophysiologic responses, as well as the evolutive character that may be progressive or regressive in more or less rapid sequences.

The polymorphism of brain contusions includes the dynamics of vascular and parenchymatous alterations and depends on their extent and severity, which in turn are related to the amplitude of the kinetic energy and the relationship between the site of the impact and the topography of the lesions (Ommaya *et al.* 1971, Stålhammar in this volume).

The evolutive character seems to be primarily determined by vascular, precapillary and capillary alterations, which by vasodilatation and vasoparalysis – with or without blood extravasations or by thrombosis – induce a degree of hypoxia and finally cause parenchymatous lesions.

Types of Brain Contusions

According to the *gravity criterion*, three types may be differentiated, namely:

Minor contusion, in which only vascular alterations are practically involved, as a rule reversible and not sufficient to induce persistent parenchymatous damages. When nevertheless these latter exist and become permanent, they can be functionally compensated.

Medium contusion, in which more ample vascular alterations exist, apt to induce evolutive parenchymatous lesions by hypoxia that may be partly compensated or may become persistent ischemic lesions finally forming microscars.

Severe contusion is the vasculo-parenchymatous response to powerful traumatic forces. The vascular and parenchymatous deteriorations succeeding rapidly one another are no longer reversible, nor can they be compensated, and they finally lead to the formation of a posttraumatic encephalopathy.

According to the *criterion of mechanism* of production there are numerous types that can be divided into two main groups, namely by cranial impact and by extracranial impact. To these may be added the rare, special type of brain contusion without any impact, the so-called 'whiplash brain injury', which will not be discussed here, as it belongs rather to physiopathology and influences the contusional morphopathology only slightly.

Macroscopic Pathology

Contusions may be recognized with the naked eye on the surface of the brain either by areas of vasodilatation with or without extravasations, or by true subpial blood effusions circumscribed or in continuity. On coronal slicing, one can see isolated or coalescent petechial hemorrhages or diffuse blood effusions in the superficial or/and in deep structures.

The macroscopic aspects of brain contusion may be evaluated by a topographic or by a gravity criterion.

According to the *topographic* criterion, or rather the site and the extent of the lesions, there are two main categories, namely:

Diffuse contusions, actually involving both hemispheres (Fig. 1), sometimes with predominance of one of them.

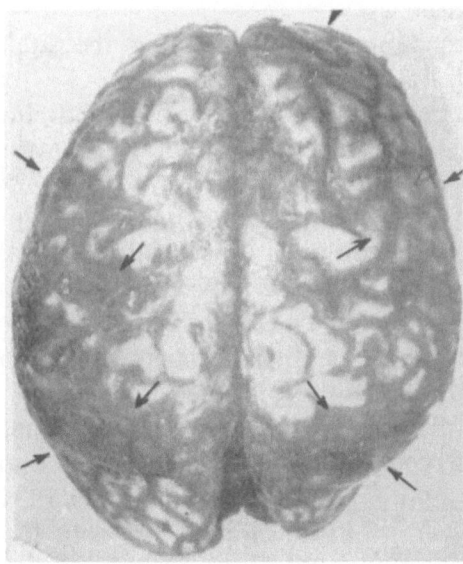

Fig. 1. Diffuse brain contusion involving both hemispheres

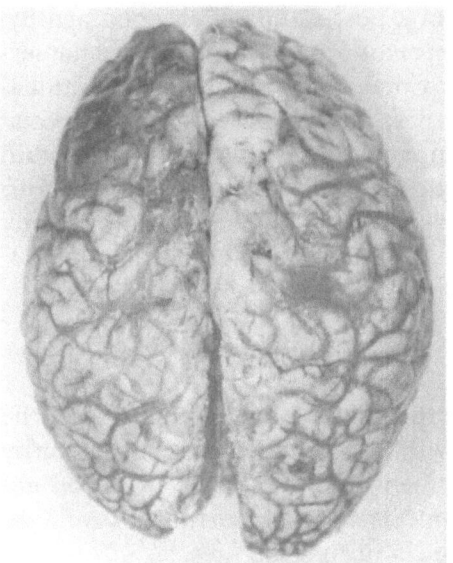 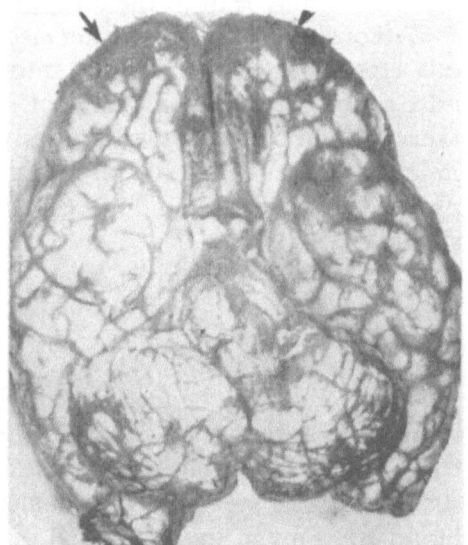

Fig. 2. Limited brain contusion involv-   Fig. 3. Limited brain contusion involving the left frontal lobe   ing both frontal poles

Limited contusions include several types, as follows

(a) Contusions involving one cerebral lobe, usually the frontal of one side
   (Fig. 2) or bilaterally (Fig. 3). Contusions involving another single lobe
   are more rarely seen.
(b) Focal contusions in an area underlying the site of the impact or
   a depressed skull fracture (Fig. 4).
(c) Contusions in some preferential areas, the most frequent and important
   being the well known fronto-orbito-temporal contusions (Fig. 5), which
   has also a corresponding clinical syndrome well delineated by
   McLaurin and Helmer (1965). Sometimes contusion is limited to the
   orbital surface of the frontal lobe associated with a small contusional
   area in the pole of the temporal lobe (Fig. 6).
(d) A more rarely encountered type is the temporo-diencephalic or the hypothalamic contusion frequently associated with complex fractures of the
   base of the skull (Figs. 7a and b), in some cases also involving the pituitary
   stalk and/or gland (Adams 1969, 1975, Hesselmann, Zülch, 1967).
(e) To these forms may be added the "contre-coup" contusions that appear
   in an area opposite to that underlying the site of the impact (Fig. 8) and
   the heterotopic contusions developing in a zone located, regardless of
   the direction of the impact forces, or not quite to a symmetrical area, as
   in the "contre-coup" lesions.

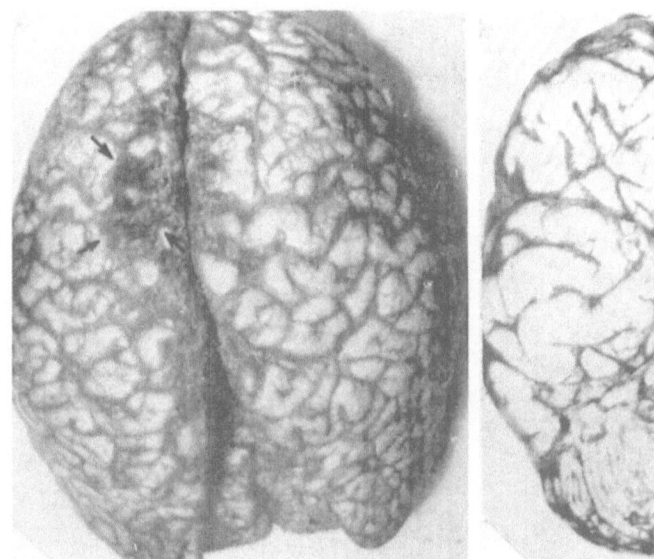

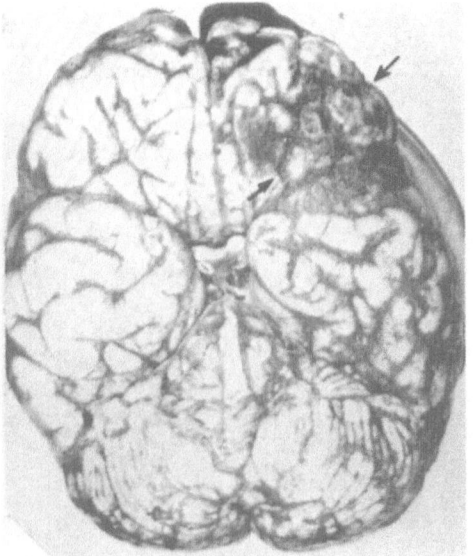

Fig. 4. Focal brain contusion (arrows) underlying the site of the impact

Fig. 5. Fronto-orbito-temporal contusion (arrows) on the left cerebral hemisphere

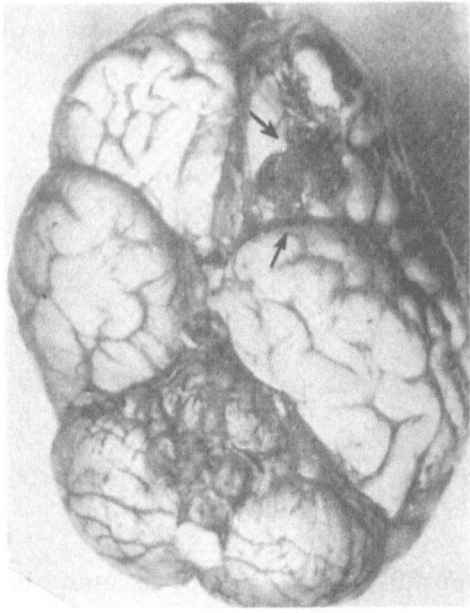

Fig. 6. Contusion limited to the orbital surface of the frontal lobe and in the pole of the left temporal lobe (arrows)

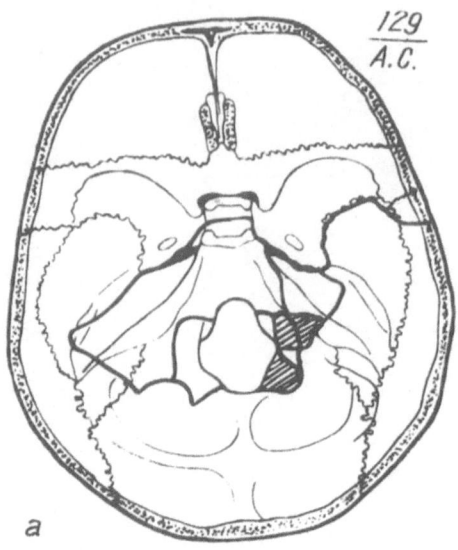

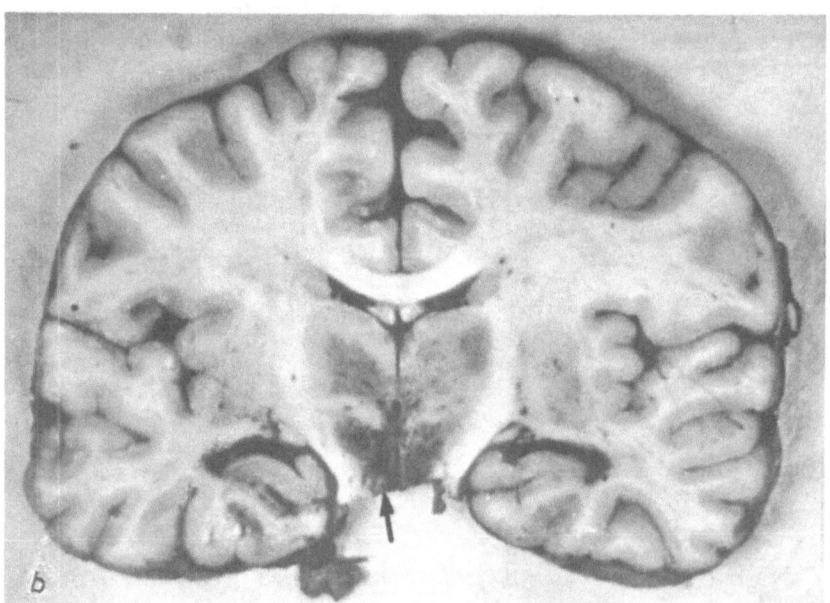

Fig. 7. Complex fracture of the base of the skull (a) producing hypothalamic contusion (b)

It is obvious that most of the limited, circumscribed contusions, and especially those over a more extensive area cannot be strictly 'limited' but are predominant in these areas, the remaining brain being more or less contused.

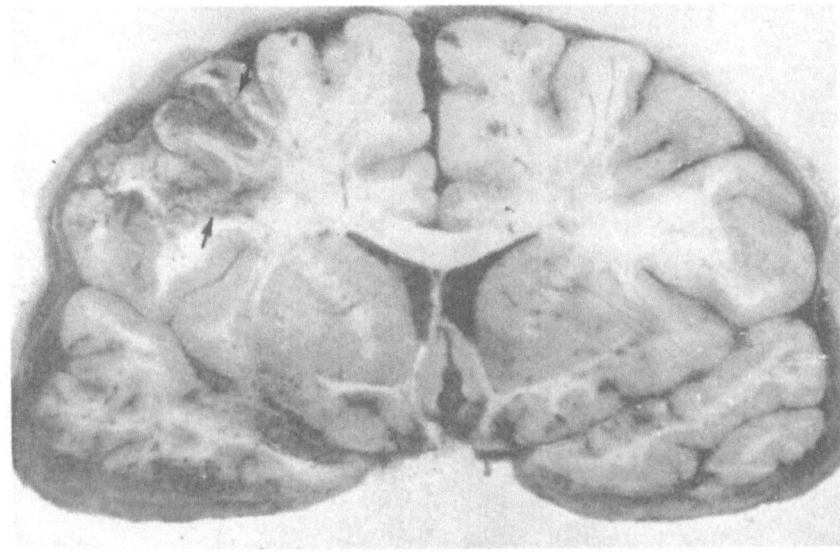

Fig. 8. Extensive areas of contusion in the left cerebral hemisphere on the side opposite to the impact ("contre-coup mechanism")

According to the *gravity criterion* (see page 30) one can again distinguish three main types of contusion:

Minor contusion in which the brain has apparently a normal appearance on examination with the naked eye. With the magnifying glass one may eventually note a slight degree of subpial vascular turgescence without blood effusion. On slicing the brain, only in some cases are very small petechiae to be seen persistent after washing.

Medium contusion in which evidence may be found of constant turgescence of the leptomeningeal and subpial vessels with small hemorrhagic areas and, on slicing, constant, visible, indelible, isolated or coalescent petechiae (Fig. 9).

Severe contusion, with practically the entire brain congested, the whole superficial vascular network is very dilated and with dense, extensive and mostly coalescent subpial blood effusions, the brain taking on a violet-hortensia blue aspect and a soft, friable consistency. The coronal slices aspect is dominated by areas of coalescent or isolated blood effusions (Fig. 10).

Adams *et al.* (1980) proposed a method for quantitative evaluation of the extent and severity of brain contusions and their preferential topography in terms of various traumatic factors.

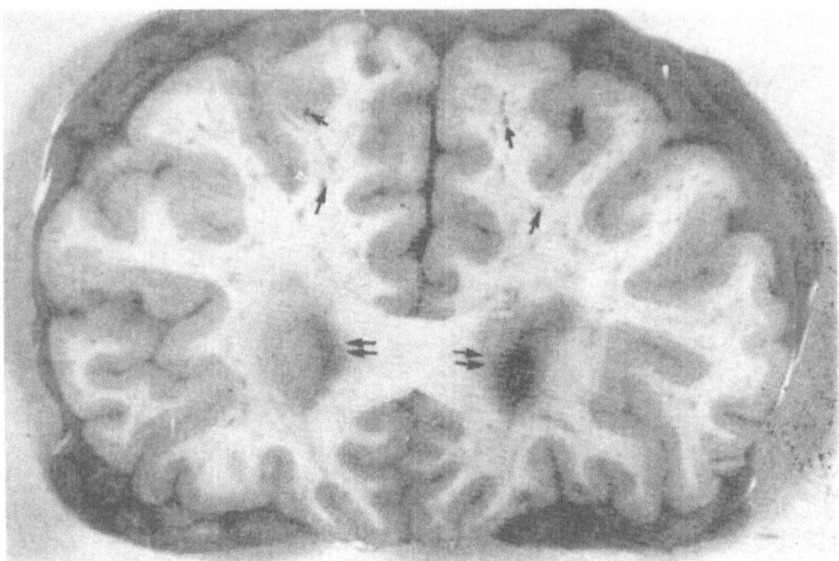

Fig. 9. Coronal slice showing diffuse, isolated and coalescent petechiae in a case of medium brain contusion

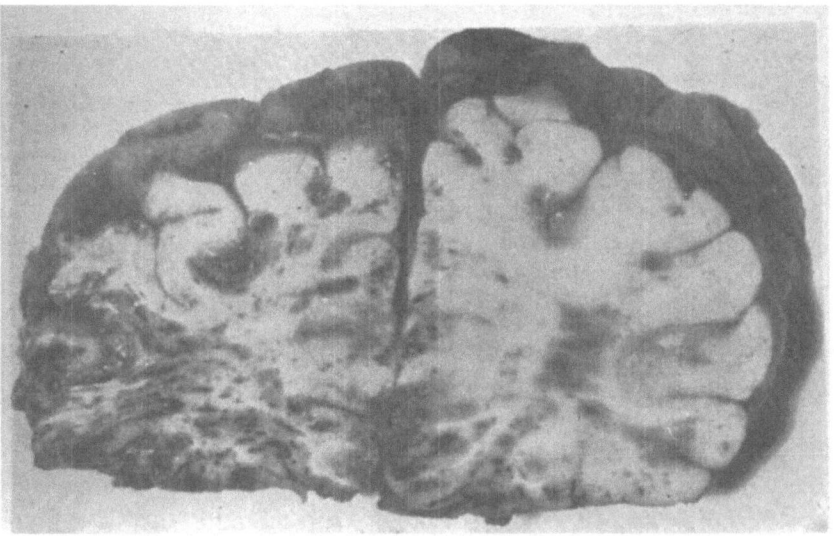

Fig. 10. Coronal slice showing dense and coalescent blood effusions in a case of severe brain contusion. Note the large subpial blood effusions

Microscopic Pathology

Microscopic analysis of the morphological alterations of brain contusion shows marked polymorphism involving, to a varied extent, all of the cerebral histological components, *i.e.* vessels, cells and fibers. The extent

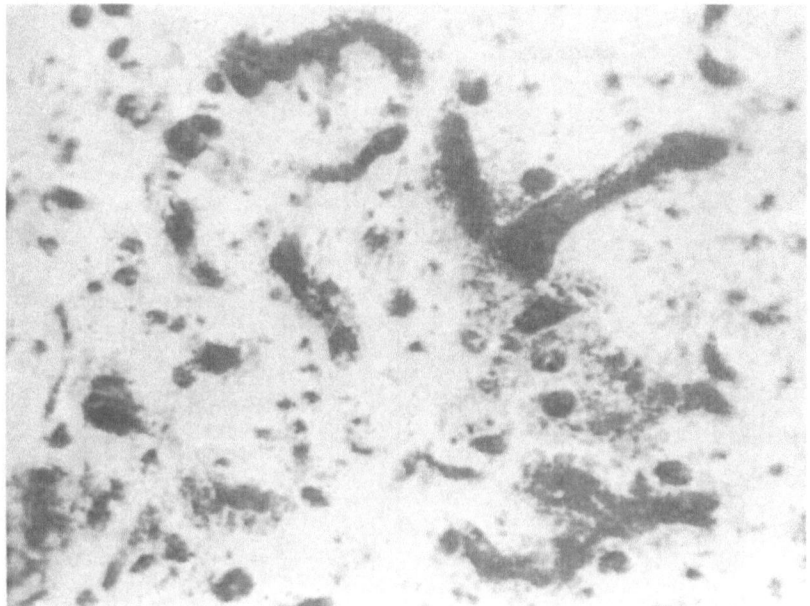

Fig. 11. Capillary and precapillary paralysis with small extravasations (minor contusion of the brain)

and the sequence of the changes is not fully known. Especially in relation to the studies of Scheincker (1944, 1945, 1947), it appears that the primary disturbances are vascular, which induce secondary parenchymatous lesions by hypoxia. It is difficult, if not impossible to reconstitute the dynamics of all of these processes, and thus, for the sake of systematisation, we shall discuss successively the vascular, the cellular and the fiber alterations.

Vascular alterations appear initially to be mainly vasodilatation, co-existing with or followed by capillary and precapillary paralysis (Fig. 11). These initial alterations have at least two consequences due to slowing of the blood flow and changes in the endothelial permeability: (1) Blood extravasations over more or less extensive areas and of various density, depending on the degree of contusion. Thus, in minor contusion the extravasations are confined and rather isolated, whereas in medium contusions the vasoparalysis and blood effusions are more extensive and more dense (Fig. 12), with an obvious tendency to coalescence, in some cases forming blood spots or even micro-hemorrhages (Fig. 13). In the severe forms of brain contusions the density of blood effusions become dominant (Fig. 14) and their coalescence can lead to the development of a true intracerebral hematoma (Fig. 15). In such cases another vascular disturbance appears: (2) Thrombosis which, in terms of the caliber of the vessels

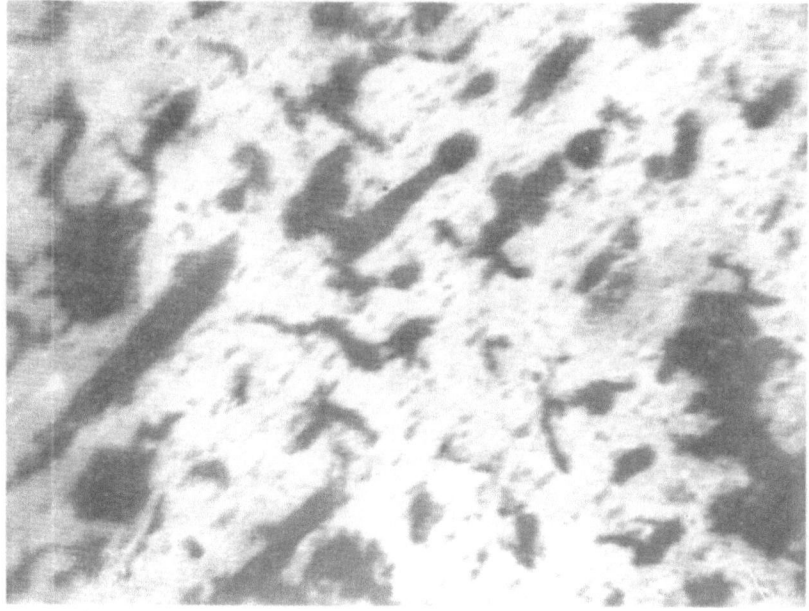

Fig. 12. Vasoparalysis with more dense and more extensive extravasations
(medium brain contusion)

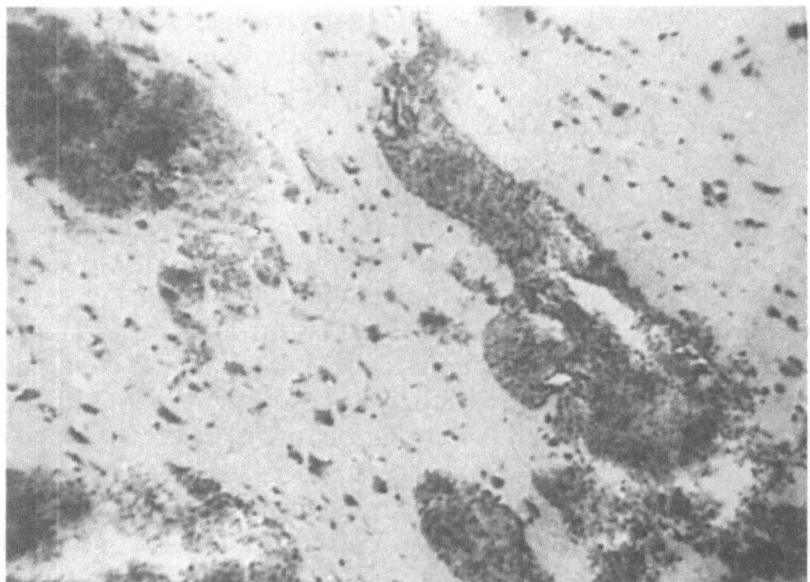

Fig. 13. Coalescence of small blood extravasations forming microhemorrhages
(medium brain contusion)

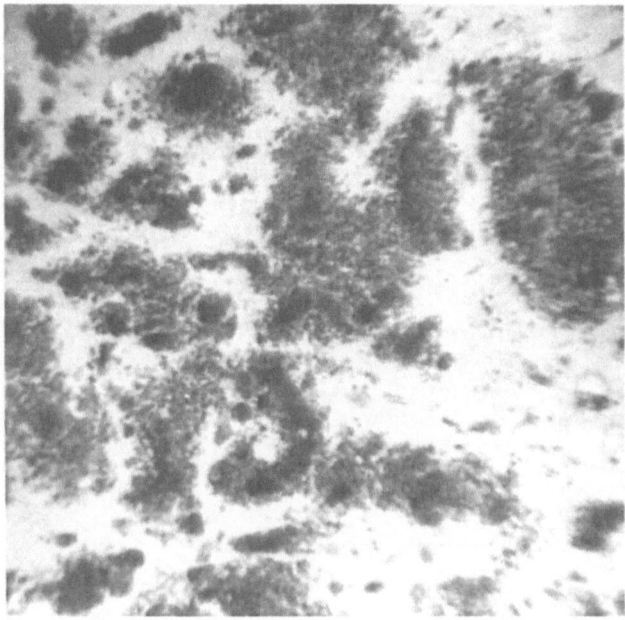

Fig. 14. Marked density and coalescence of blood effusions (severe brain contusion)

involved, produces areas of necrosis (Fig. 16), in which the development of an intraparenchymatous hematoma is facilitated. This appears to be the reason for large blood effusions and also for islands of histologically damaged cerebral tissue. In this way, the intraparenchymatous hematomas appear to be a secondary traumatic cerebral effect.

Cellular lesions concern both the neuronal and the various types of glial elements.

Neuronal lesions have been the subject of earlier studies: see Adams (1975), Arseni and Oprescu (1972, 1984). There is little doubt that the neuronal damage has an ischemic pathogenesis due to a vascular traumatic disturbance (capillary and precapillary paralysis), blood extravasations and also micro-thrombosis. Some authors also consider the ischemic areas to be due to obstruction of certain major arteries, an opinion that we feel has no direct connection with the true contusional cerebral effects. It is also necessary to mention that some recent studies attempt to correlate the neuronal lesions with raised intracranial pressure induced by several factors and mainly by cerebral edema.

In the cerebral cortex chromatolysis and hyperchromia are unspecific (Fig. 17). In severe contusions the neurons take on a vacuolar or even an areolar aspect (Fig. 18). Some authors (Rand and Courville 1946) described

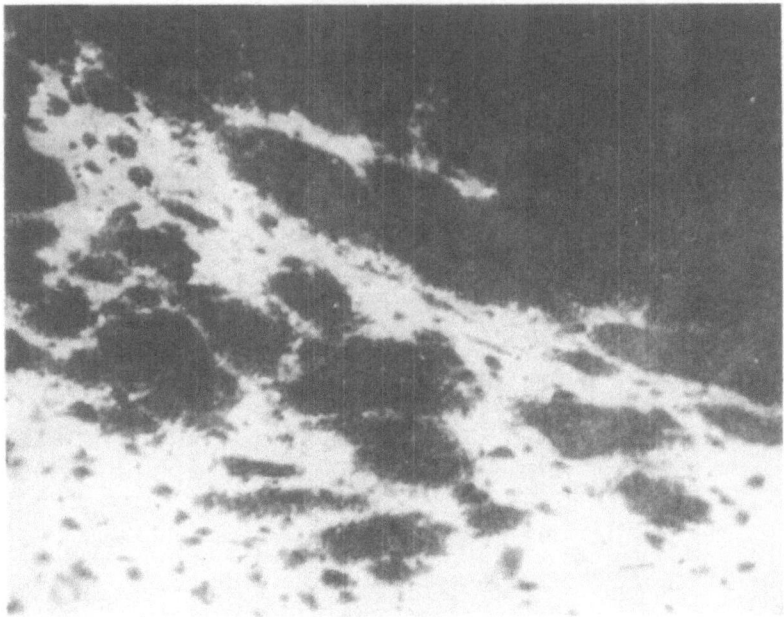

Fig. 15. Coalescence of dense blood effusions resulting in a true intraparenchymatous hematoma

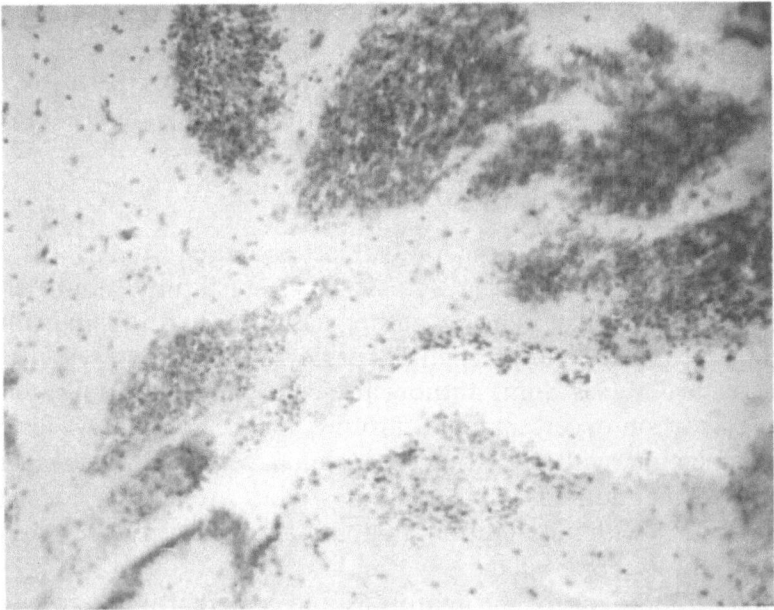

Fig. 16. Thrombosis producing necrosis and liquefaction of the cerebral parenchyma (severe brain contusion)

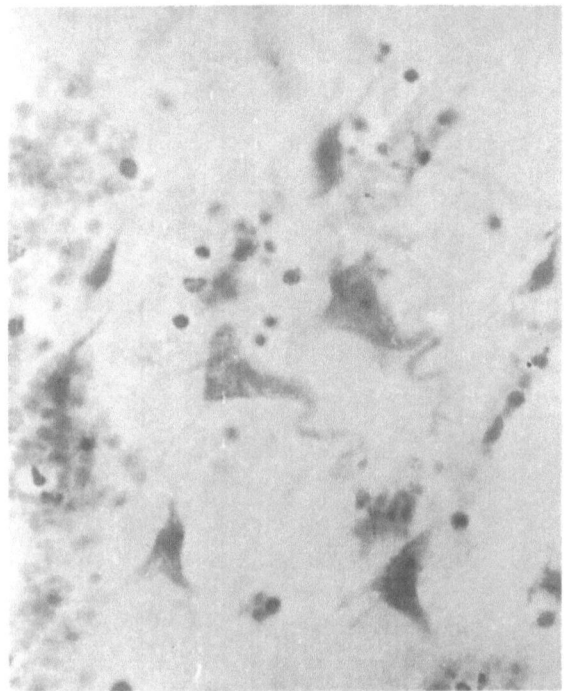

Fig. 17. Neuronal lesions (chromatolysis, hyperchromia and spiral deformation of
the apical dendrites) in brain contusion

alterations in the Golgi apparatus with a tendency to reabsorption. In the
deep structures of the brain, neuronal lesions appear only in severe contusions, taking on quantitative aspects of cellular depopulation or even of
'neuronal desert'. In addition, rather as a nonspecific phenomenon encountered also in a series of other nontraumatic lesions of the brain, a rarefaction of the Purkinje cells in the cerebellar cortex appears almost constantly,
the remaining cells being apparently intact or showing a degree of
chromatolysis.

Glial lesions are constant and fairly well known, owing especially to the
studies of Nevin (1967), Oppenheimer (1968), Clark (1974), Adams (1975).
The most significant alterations appear to be those of the microglia, mostly
with hypertrophic forms in the cerebral hemispheres, the cerebellum and in
the brain stem. In cases with prolonged survival, Adams and Graham
(1972), Adams (1975) reported on the 'grape-like' distribution, or 'microglial
stars', aspects observed also in some of our cases. Astrocytic reaction was
observed by us, both in recently injured patients as hypertrophy of various
degree, and, in cases with prolonged survival, when aspects of clasmaden-
drosis are added. Oligodendroglia appear to be less reactive and we

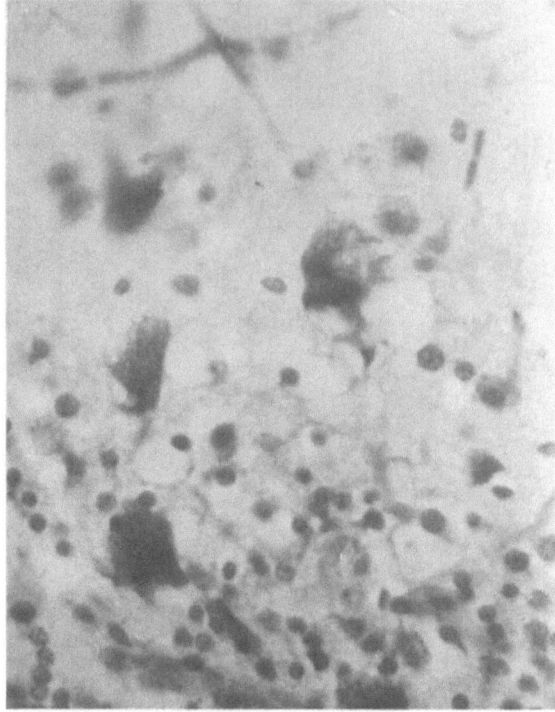

Fig. 18. Neuronal lesions (vacuolar and hyperchromia) in severe brain contusion

only found evidence of diminishing density, which did not appear to be significant.

Fiber lesions have been amply discussed, but controversies still remain. In 1928 Ramon y Cajal described in experimental material varicose and fragmentation aspects, up to granulation of the myelin sheaths. Similar findings were reported in human material by Rand and Courville (1945). More recent valuable studies are those of Strich (1956, 1961, 1969, 1970), Oppenheimer (1968), Adams (1975, 1977) upon which our knowledge is nowadays based.

One of the subjects discussed is that of the presumptive relationship between fiber alterations and posttraumatic raised intracranial pressure. A second topic would be to what extent these alterations are due directly to the primary impact effect, or to the secondary induced hypoxia. A third question is the relation of the prolonged post-impact survival which passes through several stages due to the intensive care programme.

It now seems possible to assert that fiber lesions may develop without being connected to any degree of raised intracranial pressure and therefore induced by friction phenomena in certain areas of the hemispheric white substance. As a rule, fiber lesions have a slow evolutive character which could be divided into at least three stages. A first stage, up to 14 days after the injury, would be that of axonal retraction balls, rather rapidly followed by a microglial and astrocytic reaction as described in the foregoing passages. A second stage (within days 14 to 60), and a third one, up to several months, succeeding without transition are characterized by progressive Wallerian degeneration, clearly visualized on brain slices stained by the Marchi and the Spielmayer methods.

Late Cerebral Effects

General Remarks

As was mentioned from the beginning, this discussion concerns the late effects following hemispheric contusions, the prevalent brain stem contusions being beyond the limits of our subject.

Among the effects developing late after hemispheric contusions, especially severe ones, differentiation should be made between the following groups:

1. Some of the late posttraumatic effects have a non-evolutive character and, hence, may be listed as sequelae after head injuries. Thus there are some forms of aphasia, some cases of hemiplegia, more rarely certain hyperkinetic syndromes and – again exceeding the limits of our subject – certain lesions of cranial nerves.

2. The most frequent post-contusional effects have, however, an evolutive character. From the nosologic point of view they form the group of posttraumatic encephalopathies, approximately in the same way in which meningo-cerebral scars represent the late preferential effect after brain lacerations.

3. In the encephalopathies developing within a shorter or longer interval after the primary brain contusion or after repeated head injuries, it is not always easy to distinguish accurately between specific post-contusional hemispheric lesions and those pre-existing or coexisting and possibly due to certain nontraumatic factors. More important among the latter are alcoholism, cerebral arteriosclerosis, repetitive cerebral microembolisms and cerebral hypoxia induced by acute or chronic heart or lung diseases. This group includes cases with clinically compensated hydrocephalus (Walker, Caveness and Critchley 1969; Unterharnscheidt and Sellier 1966; Adams 1975; Jennett and Teasdale 1981).

4. After prolonged coma or prolonged vegetative state a particular "protracted form of posttraumatic-encephalopathy" may develop (Jellinger and Seitelberger 1970).

Posttraumatic Encephalopathies

The term posttraumatic encephalopathy includes a group of anatomico-clinical syndromes that may develop at various intervals, ranging from a few weeks up to several months or even years after a primary contusional brain lesion. A particular form of this group, developing after repeated brain injuries and known as the "punch-drunk syndrome" will be discussed separately.

Macroscopic Pathology

In many cases of posttraumatic encephalopathy the brain is apparently normal to the naked eye or there is a slight asymmetry in the size of the two hemispheres. As a rule, arteriosclerosis especially of the vessels at the base of the brain may be observed. In some areas the leptomeninges may appear thickened with a certain degree of opalescence. In these areas the consistency of the cerebral parenchyma may be more or less hard. In rather few cases, zones of microgyria may be noted.

On coronal slices there almost constantly exists a degree of asymmetrical hydrocephalus. In the cerebral parenchyma, rarer or denser microscars and sometimes infarction areas may be seen with the naked eye, or with a magnifying glass. In a few patients small cavities with a microcystic aspect can be observed in the white matter. Marked reduction in the size of the corpus callosum, especially in its caudal segments, has also been reported.

Microscopic Pathology

The dominant alterations in posttraumatic encephalopathy occur in the white matter and smaller or minor ones in the cerebral cortex.

In the *white matter* of the cerebral hemispheres marked and, as a rule, diffuse demyelinization is observed, especially around the vessels. Various degree of myelin deterioration and axonal interruptions have been encountered, but usually sparing the association fibers, the so-called 'U' fibers. According to Strich (1956, 1961) and Nevin (1967) these aspects represent a 'secondary degeneration', probably due to the traumatic forces of distortion and friction. In the opinion of other authors, these would be the consequence of cerebral hypoxia induced by vasomotor disturbances or by cerebral edema, which would give a satisfactory explanation for their late onset. In contrast with some authors, who note a weak or moderate glial reaction, many of our cases presented rather intense proliferation, especially astrocytic, sometimes with hypertrophic forms.

In the cerebral cortex the neuronal lesions are mild, showing only a few aspects of lipofuscin accumulation.

In the cerebellar cortex the non-specific rarefaction of the Purkinje cells was noted.

The lesions of the brain stem have been amply studied, but are outwith the limits of our subject.

Punch-drunk Encephalopathy

This term refers to the late cerebral effects of repeated injuries to the head, especially in boxers, but, in fact, this anatomico-clinical syndrome developing after repeated traumas occurs also in other sports such as rugby, baseball et cetera, and even after repeated non-sport traumas.

Some authors maintain that this is a "pure" form of posttraumatic encephalopathy, but this seems questionable because many boxers are alcoholics, some excessively drugged, and both the boxers and other athletes acquire, owing to their hard and prolonged training, certain degrees of heart diseases which may cause in time progressive cerebral hypoxia. Therefore it may be assumed that the late cerebral lesions are nevertheless the consequence of several factors, repeated trauma being the predominant one.

Our present knowledge is based especially on the data furnished by Unterharnscheidt in a series of studies summarized in 1975, by Brandenburg and Hallervorden (1954), Graham and Ule (1957), Payne (1968), Corselis et al. (1973).

Macroscopic Pathology

In some cases decrease in the weight of the brain with dilatation of the entire ventricular system was reported, as well as overall thinning of the corpus callosum and septal anomalies of cavitary type. On coronal slicing, narrowing of the cortex, mini- or microscars were observed with a magnifying glass.

Microscopic Pathology

The cortex of the cerebral hemispheres reveal neuronal rarefaction and a less systematized stratification. Aspects of 'status spongiosus' can develop, especially in the occipital cortex. More marked than in other encephalopathies, gliosis of astrocytic-fibrillary type may be detected. In the thalamus, the astrocytic glial reaction is prevalently of plasmatic type.

The white matter presents areas or zones of demyelinization.

In the cerebellar cortex the same non-specific rarefaction of the Purkinje cells and reduced thickness of the granular layer were noted.

Corselis et al. (1973) comparing the punch-drunk encephalopathy brain with that of other posttraumatic encephalopathies, showed the more frequent presence in the former, of septal anomalies and the virtual absence of senile plaques in the cases with demential syndromes, but with neurofibrillary tangles in both cases.

Cerebral Hemispheric Lacerations

Lacerations of the brain may be defined as primary, immediate post-traumatic effects with a destructive character, therefore necessarily involving discontinuity by disruption of cerebral parenchymatous structures. In some cases the damage may be limited to the cortex, but as a rule the cortico-subcortical structures are involved more or less deeply. Among the deep hemispheric structures more frequently implicated, are the temporo-frontal and temporo-rhinencephalic, and rather rarely the basal and sub-thalamic nuclei or the hypothalamo-hypophyseal axis.

The concept of cerebral laceration as a separate neuropathological (and clinical) entity is not very old. The first studies on brain lacerations were published by Columella (1973), de Pian (1967), Rusu (1967). In the French literature, the term "attrition" is equivalent to laceration and was reported in connection with subdural hematomas: Arseni and Oprescu (1977, 1984), Vigouroux and Guillermain (1981), Jennett and Teasdale (1981). Nevertheless many aspects remain questionable and have to be clarified in the future.

Laceration very rarely occurs as a single lesion in the damaged brain. As a rule, it is associated with other primary, secondary or subsequent traumatic effects. This is inevitable because the kinetic energy able to induce laceration, concomitantly produces a degree of contusion and of brain edema. Under certain conditions, disruption of the parenchyma also implies vascular lesions, giving rise to blood effusions *i.e.* intra- or extracerebral hematomas.

Immediate Cerebral Effects

Types of Brain Lacerations

There are two main types of primary lacerations to which one can add a much rarer type of secondary laceration of the brain (Arseni and Oprescu 1977).

1. *Direct cerebral laceration* represents a destructive lesion of the cerebral parenchyma caused by intracranial penetration either of a foreign body (missile, side arms, metallic fragments projected with high speed) or of a penetrating craniocerebral wound.

2. *Indirect cerebral laceration* is a particular form more recently recognized. It is the consequence of the difference in the inertia of the rigid, irregular structures of the dural-skull container and of the vascular and parenchymatous content resulting in frictions between the two types of structures, possibly or inevitably leading to brain lacerations (see the chapter by Nakamura in this volume).

As in brain contusions, the most frequent mechanisms are abrupt deceleration and abrupt acceleration. In *linear acceleration* at low or medium velocity, disruption of the brain does not generally occur (contusion being more frequent), but at high velocities cranio-cerebral wounds appear, in which brain laceration is one component. In the *rotatory acceleration*, distortion of the skull and of the brain may produce lacerations, usually cortical. As a rule, indirect lacerations do not imply skull fractures.

3. *Secondary cerebral lacerations* are produced by cortical disruption either by a traumatic or by a primary intraparenchymatous hematoma with a rapid expansive character.

Secondary circumscribed lacerations subjacent to the area of impact may exist due to a rapid and powerful cranial inflection–deflection effect, with or without skull fractures, but they are rather rare.

Macroscopic Pathology

Any type of brain laceration may readily be recognized with the naked eye by its disruptive, destructive character, as shown in its definition. The damage is exclusively located in a more or less circumscribed area, therefore diffuse forms do not exist.

As a rule, brain laceration develops as a *single lesion*, with a preferential topography in terms of its mechanisms. *Multiple lacerations* are exceptional and produced by complex mechanisms (direct impact plus contre-coup or powerful rotatory acceleration) and in repeated head injuries at short intervals.

The simplest form of brain laceration is the *cortical circumscribed type*, due to an abrupt inflection–deflection phenomenon in an area of the skull. If more extensive, such a mechanism may produce a cortico-subcortical disruption (Fig. 19).

In *direct lacerations*, parenchymatous disruption virtually corresponds to the topography of the impact, and is more or less extensive and deep according to the physical properties of the damaging agent, especially its velocity. Usually these lesions are cortico-subcortical involving two or three gyri on the surface and 2–4 cm in depth (Fig. 20). If the applied force is greater, associated contusion and perifocal edema result in protrusion of the lacerated, necrotic cerebral parenchyma (Fig. 21).

In *indirect lacerations* the brain lesion does not constantly correspond to the site of the impact. The temporal lobe is preferentially implicated in the destructive process. Rusu (1972) convincingly described the existence of "indirect circumscribed and localized laceration" in a temporo-basal area, but this type is rare. The cerebral areas more often involved are the temporal lobe and/or the inferior, orbital aspect of the frontal lobe (Fig. 22).

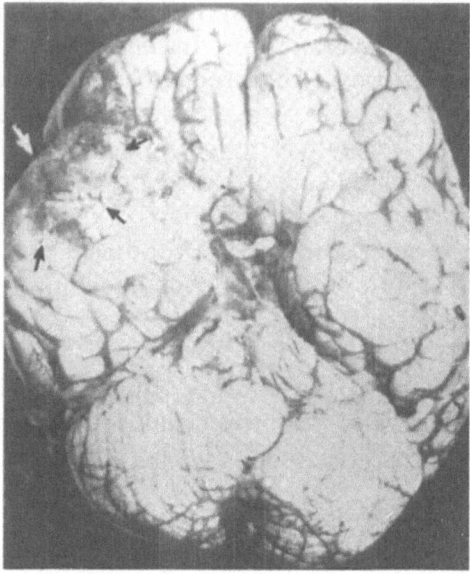

Fig. 19. Cortico-subcortical circumscribed brain laceration due to abrupt inflection–deflection mechanism

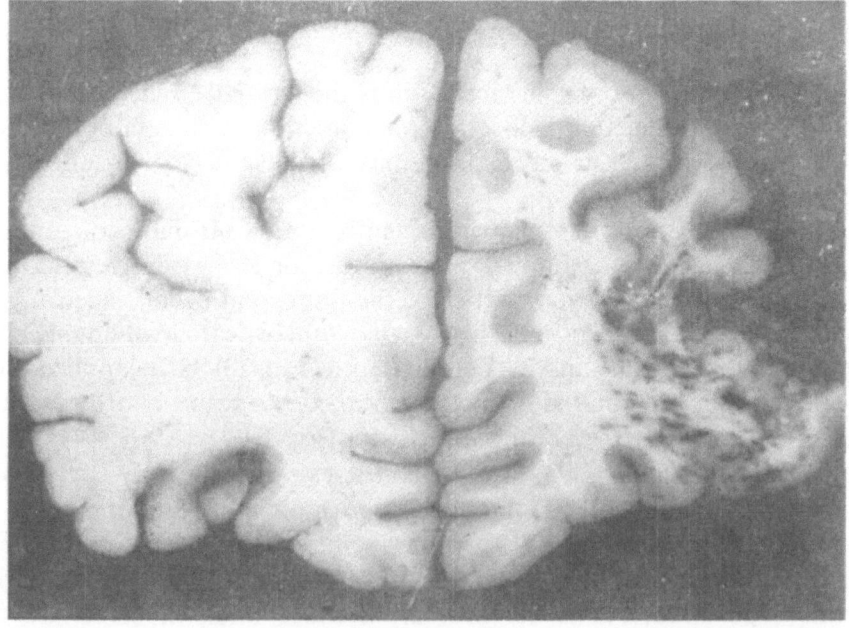

Fig. 20. Direct brain laceration involving three gyri on the left hemisphere

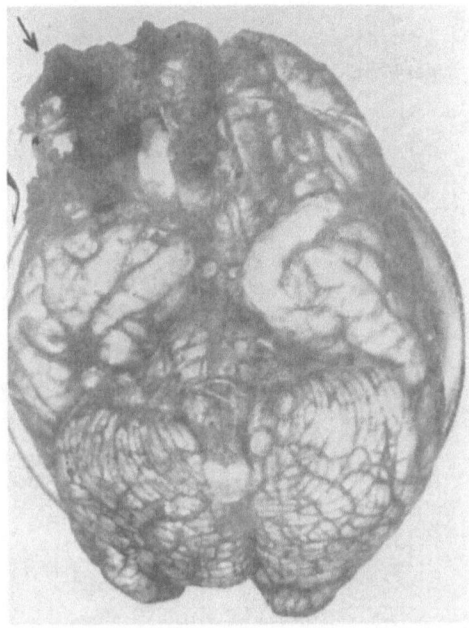

Fig. 21. Extrusion (arrow) of the lacerated, necrotic cerebral parenchyma of the right frontal lobe in a case of severe direct laceration

When both of these areas are implicated, the so-called "temporo-orbital laceration" results (Fig. 23). It may be relatively superficial (cortico-subcortical) or may include the deep temporal structures *i.e.* the amygdaloid and the hippocampal complex. Indirect brain lacerations rarely involve the temporal structures predominantly, but rather uni- or bilaterally the frontal lobes or poles especially in cases with occipital impact. There are also cortico-subcortical lacerations with other topography, but they are exceptional.

A particular form of brain laceration is that produced by certain severe head injuries with dehiscent basal skull fractures usually running rostro-caudally. Dehiscence of the bone induces laceration of the basal dura-mater, as well as of the diencephalic and temporo-medial structures, sometimes even disrupting the corpus callosum (Fig. 24).

Microscopic Pathology

In the actual laceration area the histological components of the cerebral parenchyma cannot be clearly recognized, only a few deteriorated elements are identifiable.

Light microscopy almost exclusively reveals a mixture of islands of parenchymatous necrosis and of large areas of blood effusions (Fig. 25). In

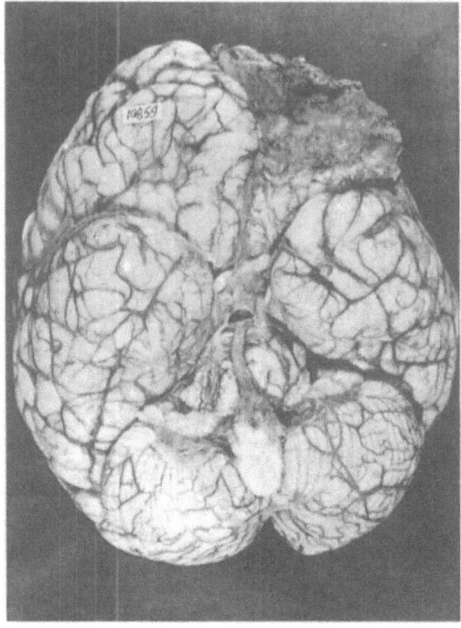 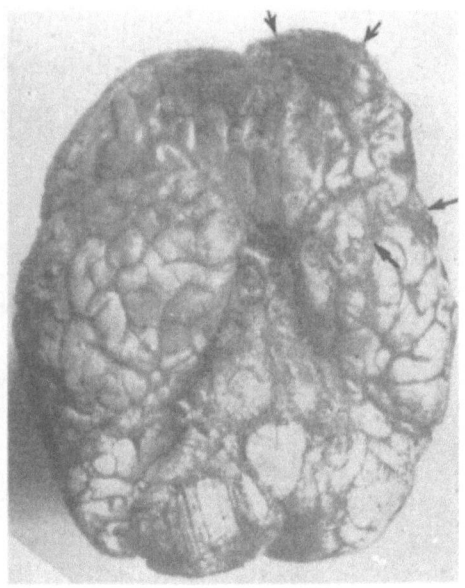

Fig. 22. Indirect brain laceration circumscribed to the orbital area of the left
frontal lobe

Fig. 23. Indirect brain laceration circumscribed in the temporo-orbital
areas of the left hemisphere

this background of necrosis, some intact capillaries can be seen, but more
frequently deteriorated, and in many instances as blood micro-extravasa-
tions. In the immediate peripheral areas and those further from the laceration zone, contusion and cerebral edema are almost constant. Similarly,
evidence is found of rare, altered or deteriorated neurons as real "neuronal
shadows" (Fig. 26), rare glial cells and in some cases granulous bodies are
the result of microglial phagocytic activity.

Electron microscopy does not appear to have provided new data concerning the disintegration of the cerebral parenchyma in which necrosis is
predominant, as in light microscopy. It is important to note the studies of
Testa, Bollini and Columella (1970) on vascular alterations in the laceration
areas. Thus, in the first seven days, significant capillary endothelial lesions
can be observed, developing up to necrosis and sometimes to subsequent
cellular disappearance, but without implication of the basal membrane. In
the later phases up to three months, the basal membrane appears virtually
disintegrated and the capillary endothelium flattened. In the sequences of
these lesions, small discontinuities occur, permitting leakage of plasma,
a phenomenon designated by Columella (1973) as "plasmarrhage". In

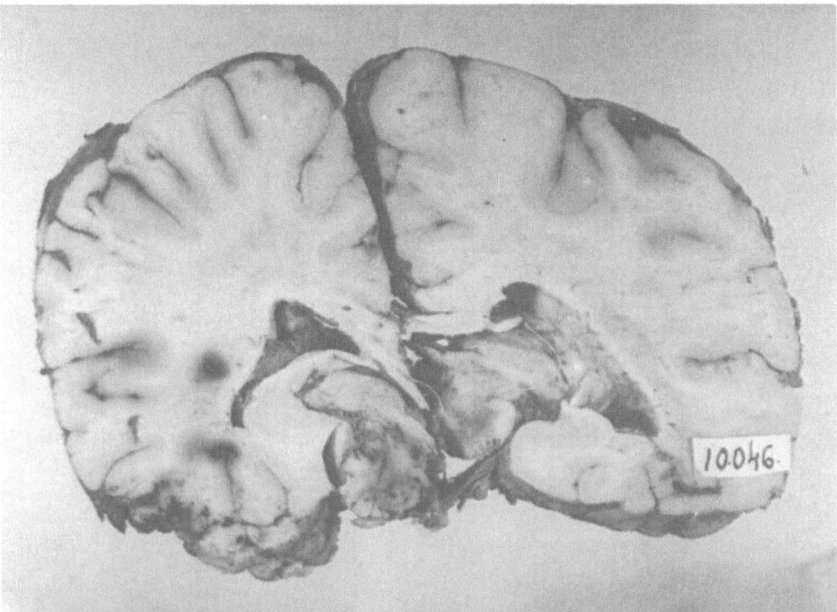

Fig. 24. Temporo-diencephalic and callosal disruptions

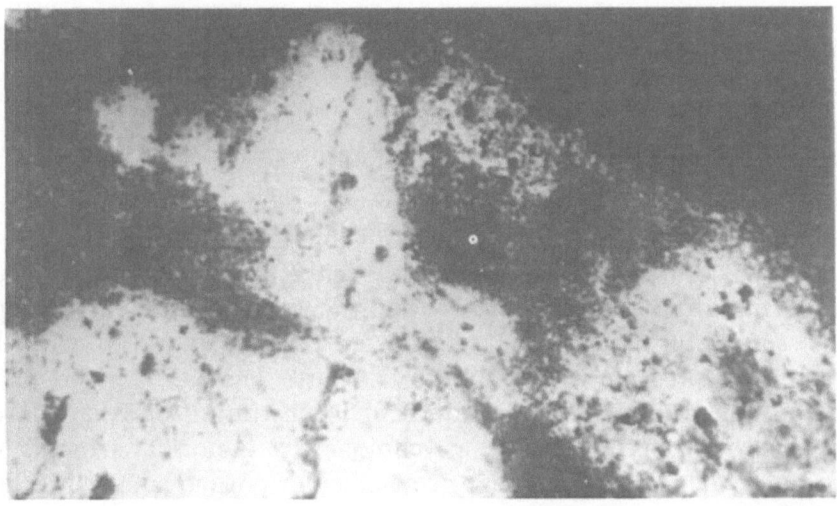

Fig. 25. Islands of parenchymatous necrosis and large areas of blood effusions in brain lacerations

a final stage, numerous collagen boundaries appear around the capillaries, no discontinuity is observed in the capillary walls and the basal membrane seems to be restored.

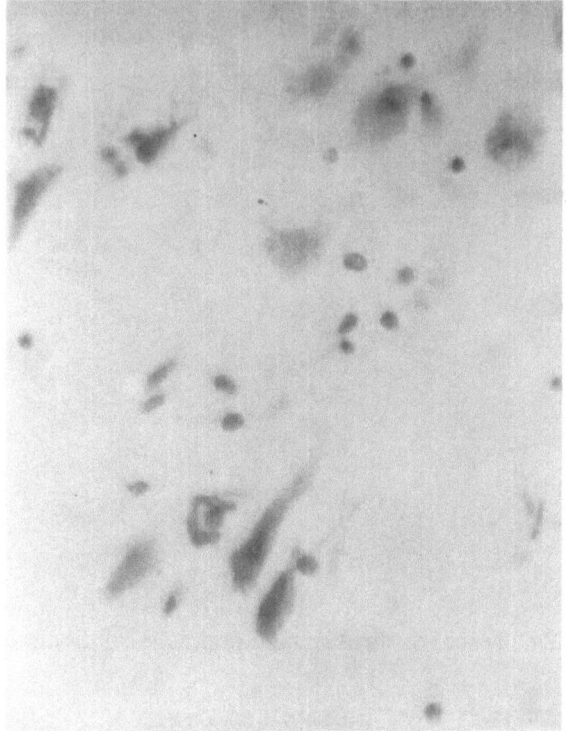

Fig. 26. "Neuronal shadows" around an area of brain laceration

Lacerations as Component of Cranio-cerebral Wounds

By definition, any kind of cranio-cerebral wound (tangential, by rico-chet, blunt, penetrating blind or transfixing) has a brain laceration compon-ent. The type of brain laceration is very varied in terms of several factors, the most important being the nature and the velocity of the impact agent.

The simplest form of brain laceration is a cranio-cerebral wound, which presumably represents the only type of "pure" laceration without asso-ciated lesions, caused by slow endocranial penetration of a hard foreign body. This may occur in some psychologically abnormal patients with a suicidal tendency. A fine linear or canalicular cortico-subcortical lacer-ation is produced, around which there are no associated lesions, but there may be exceptionally a weak contusional or edematous reaction.

Laceration prototype, as a component of a cranio-cerebral wound, is produced by a depressed cranial fracture when a bone fragment tears the dura-mater and penetrates the brain. Brain disruption may be only cortical, but is usually cortico-subcortical and frequently surrounded by a degree of perifocal contusion and edema. In the laceration focus, scalp fragments, hair or other foreign bodies may be found penetrating together with the impacting agent.

The cranio-cerebral wounds by high velocity missiles are most frequent in war-time and are rather rare in civilian life.

The following main types may be delineated:

The *tangential* cranio-cerebral wounds in which the projectile disrupts the scalp, fractures the skull, a bone fragment tears the dura-mater and produces laceration of the subjacent brain. The laceration is usually conical with the base towards the endocranium. Similar wounds are produced by the ricochet mechanism.

The *blind penetrating type* is the most common (50%). The bone fracture is usually orificial-like and the brain laceration has a more or less canalicular aspect. The amplitude of laceration is greater in the cortico-subcortical areas and more or less restricted in depth because the vulnerant agent velocity decreases gradually. There is a marked perifocal contusion or even necrosis and extensive brain edema.

The *true penetrating type* is characterized by severe and extensive damage, to both the skull and the brain, directly proportional to the high velocity of the missile. According to the direction of the impact forces, these wounds may be segmental, diagonal or diametrical. In all these events the laceration area is approximately canalicular, with an unequal diameter that generally increases from the site of the entry to its exit. The very high velocity forces generate shock waves into the brain with radiating propagation. This would explain the extensive areas of marked contusion at a great distance from the laceration trajectory and the magnitude of cerebral edema. These lesions with rapid, abrupt raised intracranial pressure may add to the initial skull fractures, dislocations of the cranial sutures resulting in the so-called "cranial explosion". If a laceration also involves the ventricular system, cerebro-spinal fluid and liquefied parenchymal detritus escape through the cranial fractures, and may cause severe clinical complications.

Progressive Continuous Form of Brain Lacerations

This form generally occurs in indirect brain lacerations and is considered as such when there is neither blood effusion in the laceration area, nor a secondary hematoma. Progressive brain laceration has the character of a space-occupying process. This is not due, as was initially asserted, only to contusion and perifocal cerebral edema, but to the summation of several factors not all understood as of yet. One of these would be necrosis of the lacerated brain parenchyma consequent to the vascular perturbations and lesions. To these may presumably be added the release of little known substances which may belong to the enzymatic series, *i.e.* hydrolases, oxyreductases and alkaline phosphatases, also found in the evolution of cerebral experimental scars.

Late Lesional Effects of Brain Lacerations

As for cerebral contusions, brain lacerations have the same two categories of late effects, *i.e.* the sequellar and the evolutive ones. The main evolutive late effect after a brain laceration is a meningo-cerebral scar. After severe head injuries with lacerations associated with brain contusion and edema, the late lesional effects can be a meningo-cerebral scar plus a form of encephalopathy.

Meningo-cerebral Scars

The reorganization of the lacerated brain parenchyma begins relatively early, passes through several intermediate stages to reach its final histopathological proper structure.

Both ectodermal and mesenchymal components are implicated in this process. There also appears to be an ectomesenchymal barrier whose rupture by the traumatic forces might interfere in the histogenesis of the meningo-cerebral scars, following the combination of the two components.

Within 5–7 days there appear in the laceration focus granular bodies (macrophages) resulting from proliferation and phagocytic activity, both microglial and mesenchymal, of vascular origin, with subsequent astroglial proliferation.

Following the terminal proliferative stages, fibroblastic and fibrocytic invasion of leptomeningeal origin takes place. Fibrillogenetic activity of histiocytic elements produce more or less abundant collagen and argentaffin fibers. Neoformative capillaries with a partly intact structure but frequently with thrombosis or parietal alterations, may be discerned in the early stages in the development of the meningo-cerebral scars. This neocapillary network provides a precarious supply of blood in the scar area and its periphery, thus the induced hypoxia contributes to the extent of the future scar.

Macroscopic Pathology

The forementioned intricate morphogenetic factors may produce several types of meningocerebral scars, of which the most distinct are as follows:

The *predominant meningeal scar* is a rare form involving in particular the pachymeninges. A more or less circumscribed hardelastic zone appears on

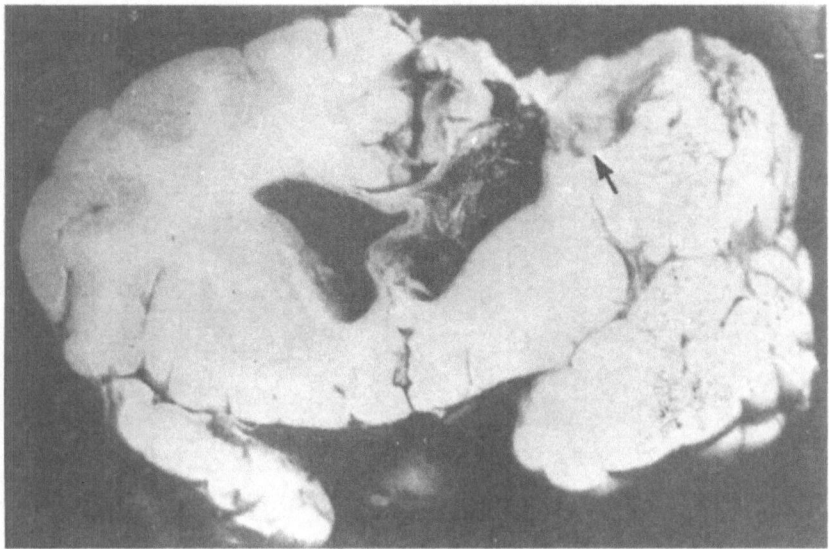

Fig. 27. Macroscopic aspect of a meningo-cerebro-ventricular scar (arrow) in the right hemisphere

the internal aspect of the dura-mater, from which fine adhesions form with the leptomeninges and with the cerebral cortex.

The *meningo-cortical scar* is the most frequent type. The leptomeninges appear thickened with a whitish-opalescent aspect, adhering to a friable yellowish-green cortex, sometimes presenting porosities and microcysts, and at the periphery a slight zone of sclerosis.

The *meningo-cerebral scar*, the so-called "en bloque cicatrix", penetrates the cerebral parenchyma deeply and as a rule to the vicinity of the ventricular wall, forming the type known as "meningo-cerebro-ventricular scar" (Fig. 27). This is a hard, irregular yellow or opalescent mass. The lepto- and the pachymeninges cannot be separated from the parenchyma, and cystic formations with a fluid, viscous or gelatinous content may be observed.

Microscopic Pathology

Certain differences exist between the microarchitectonics of the meningo-cerebral scars, more often in open head injuries, and the cerebral scars, predominant in closed head injuries (Arseni and Oprescu 1984).

The *meningo-cerebral scar* has a morphopathologic organization in which several areas may be discerned (Fig. 28): A central or mesenchymal area in which collagen fibers predominate, distributed in more or less systematised bundles, argentaffin especially towards the periphery and unequally distributed, with some fibroblastic and histiocytic infiltration, among which rare mast and plasmacells can be seen. An intermediate or

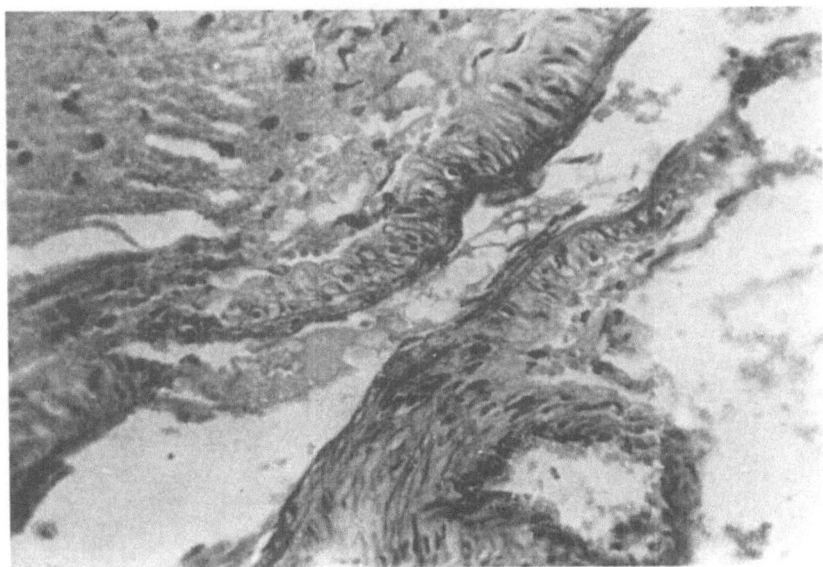

Fig. 28. Microscopic aspect of an "organized" meningo-cerebral scar

mesenchymo-glial area is found in which connective fibrous tissue elements penetrate from the central area, with few or no argentaffin fibers, frequent granular bodies (macrophages) and a microglial proliferation with few astrocytes. A peripheral or glial area dominated by microglial proliferation and with persistent elements originating from the intermediate area; hypertrophic forms and globe-like cells can be seen. More or less frequently astrocytes of fibrillary type with a perivascular distribution and some piloid or amoeboid forms may also exist. This area also presents demyelinization zones ranging from moniliform to fragmentary aspects, or complete demyelinization. There exists also an area of cerebro-cicatricial transition in which neuronal forms appear in different degenerative stages (hyperchromia, shrinking, vascuolization) and sometimes of "neuronal shadows" type.

The *cerebral scar* is, as a rule, multiple and appears as areas of circumscribed atrophy (focal cerebral scar) consequent to anoxia or precapillary blood effusion zones. There is little collagen, the dominant aspect being one of astro-microgliosis with atypical forms similar to those found in atrophic sclerosis of the brain.

Summary

The neuropathological findings of cerebral hemispheric contusions and lacerations, as described precedingly, are summarized in Table 2, following the basic classification of immediate and late cerebral effects.

References

1. Adams JH (1975) The neuropathology of head injuries. In: Vinken PJ, Bruyn EW (eds) Handbook of clinical neurology, Vol 23, Part 1, Chap 3. North Holland Publ. Co., Amsterdam, pp 378–391
2. Adams JH, Graham DI (1972) The pathology of blunt head injuries. In: Critchley M, O'Leary J, Jennett B (eds) Scientific foundation of neurology. Heinemann, London, pp 478–491
3. Adams JH, Mitchell D, Graham DI, Doyle D (1977) Diffuse brain damage of immediate impact type. Brain 100: 389–402
4. Adams JH, Scott G et al (1980) The contusion index. A quantitative approach to cerebral contusions in head injury. Neuropathol Appl Neurobiol 6: 319–324
5. Adams RD (1969) In: Walker EA, Caveness WF, Critschley M (eds) The late effects of head injury. Ch C Thomas, Springfield, Ill.
6. Arseni C, Oprescu I (1972) Traumatologia cranio-cerebrala. Editura Medicala, Bucuresti
7. Arseni C, Oprescu I (1977) The basis of a new classification of cranio-cerebral injuries. Seara Medica 6: 107–114
8. Arseni C, Oprescu I (1984) Neurotraumatologie. Editura Didactica si Pedagogica, Bucuresti
9. Brandenburg W, Hallervorden J (1954) Dementia pugilistica mit anatomischem Befund. Virch Arch (A) 325: 680–709
10. Cajal SR (1928) Degeneration and regeneration in the nervous system. Oxford University Press, London
11. Clark JM (1974) Distribution of microglial clusters in the brain after head injury. Neurosurg Psychiat 37: 463–474
12. Columella F (1973) Traumatic brain laceration as a new and independent entity in neurosurgical pathology. Neurocirugia (Chile) XXXI, 1–2: 9–26
13. Corselis JAN, Burton CJ, Freeman-Browne D (1973) The aftermath of boxing. Psychol Med 3: 270–303
14. Da Pian R et al (1967) Lacerationi cerebrali traumatische. Considerazioni su 190 casi operati. Minerva Neurochir 11: 147–153
15. Graham H, Ule G (1957) Beitrag zur Kenntnis der chronischen und cerebralen Krankheitsbilder bei Boxern (Dementia pugilistica und traumatische Boxerencephalopathie). Psychiat Neurol (Basel) 134: 261–283
16. Henn R (1989) Traumatische Veränderungen. In: Cervós-Navarro J, Ferszt R (eds) Klinische Neuropathologie. Thieme, Stuttgart New York
17. Hesselmann J, Zülch KJ (1967) Vegetative und endokrine Symptome nach traumatischer Hypothalamusschädigung. Acta Neurovegetat (Wien) 30: 251–260
18. Jellinger K, Seitelberger G (1970) Protracted posttraumatic encephalopathy: pathology, pathogenesis and clinical implications. J Neurol Sci 10: 51–94
19. Jennett I, Teasdale B (1981) Management of head injuries. F.R. Davis Company, Philadelphia

20. McLaurin RL, Helmer F (1965) The syndrome of temporal lobe contusion J Neurosurg 23: 195–304
21. Nakamura N (1990) Diffuse brain injury and brainstem dysfunction. In this volume
22. Nevin NC (1967) Neuropathological changes in the white matter following head injuries. J Neuropathol Exp Neurol 26: 77–84
23. Ommaya KA, Gennarelli AT (1975) Experimental head injury. In: Vinken PJ, Bruyn EW (eds) Handbook of clinical neurology, Vol 23, Chap 4. North Holland Publ. Co., Amsterdam, pp 67–90
24. Ommaya KA, Grubb RL, Nauman RA (1971) Coup and contre-coup injury: observations on the mechanics of visible brain injuries in the rhesus monkey. J Neurosurg 35: 503–536
25. Oppenheimer DR (1968) Microscopic lesions in the brain following head injury. J Neurol Neurosurg Psychiatry 31: 299–306
26. Payne EE (1968) Brain of boxers. Neurochirurgia (Stuttgart) 11: 173–188
27. Rand CW, Courville BC (1945) Alterations of the myelin sheats adjacent to traumatic lesions of the brain. Bull Los Angeles Neurol Soc 10: 19–34
28. Rand CW, Courville BC (1946) Histologic changes in the brain in cases of fatal injury of the head VII. Alterations in nerve cells. Arch Neurol Psychiat (Chicago) 55: 79–110
29. Rusu M (1972) La dilacération meningo-cérébrale traumatique localisée. Neurochirurgia (Stuttgart) 15: 209–217
30. Scheinker IM (1944) Vasoparalysis of the central nervous system, a characteristic vascular syndrome. Significance in the pathology of the central nervous system. Arch Neurol Psychiat (Chicago) 12: 32–43
31. Scheinker IM (1947) Cerebral swelling. Histopathology, classification and clinical significance of brain edema. J Neurosurg 4: 255–258
32. Scheinker IM (1945) Vasothrombosis in the central nervous system. A characteristic vascular syndrome caused by a prolonged state of vasoparalysis. Arch Neurol Psychiat (Chicago) 53: 171–184
33. Strich SJ (1956) Diffuse degeneration of the cerebral white matter in severe dementia following head injury. J Neurol Neurosurg Psychiatry 19: 163–185
34. Strich SJ (1961) Shearing of nerve fibers as a cause of brain damage due to head injury. Lancet ii: 443–448
35. Strich SJ (1969) The pathology of brain damage due to blunt head injury. In: Walker AE, Caveness WF, Critchley M (eds) The late effects of head injuries. Ch C Thomas, Springfield, Ill
36. Strich SJ (1970) Lesions in the cerebral hemispheres after blunt head injury. J Clin Pathol 23 [Suppl 4]: 154–165
37. Testa C, Bollini C, Columella F (1970) Cerebral capillaries in the evolution of traumatic lacerations in man: an ultrastructural study. Folia Angiographica 96: 83–89
38. Unterharnscheidt F (1975) Injuries due to boxing and other sports. In: Vinken PJ, Bruyn EW (eds) Handbook of clinical neurology, Vol 23, Part I, Chap 26. North Holland Publ. Co., Amsterdam, pp 527–593

39. Unterharnscheidt F, Sellier K (1966) Mechanics and pathomorphology of closed head injuries. In: Caveness FW, Walker EA (eds) Head Injury Conference Proceedings. Lippincott, Philadelphia, pp 321–341
40. Vigouroux PR, Guillermain P (1981) Posttraumatic hemispheric contusions and lacerations. In: Krayenbühl H *et al* (eds) Progress in neurological surgery. S Karger, Basel, pp 49–163
41. Walker EA, Caveness FW, Crichley M (eds) (1969) The late effects of head injury. Ch C Thomas, Springfield, Ill

Diffuse Brain Injury and Brainstem Dysfunction

N. NAKAMURA

Department of Neurosurgery, Tokyo Jikei University School of Medicine,
Tokyo (Japan)

With 33 Figures

Contents

Introduction

A Glasgow Neuropathologist, Adams, and the Pennsylvanian Neurosurgeons Bruce and Gennarelli introduced the terms "diffuse brain damage" and "diffuse axonal injury" on the basis of experimental data and on clinical and pathological observations (Adams *et al.* 1977, 1983, Bruce *et al.* 1978, 1981). Zimmermann, Bilaniuk and Gennarelli (1978) described the neuroradiological findings of shearing injuries of the white matter as seen on computerised tomography.

According to the engineer Holbourn (1943) at the moment of impact, a shearing force develops in the brain in response to the inertia difference of

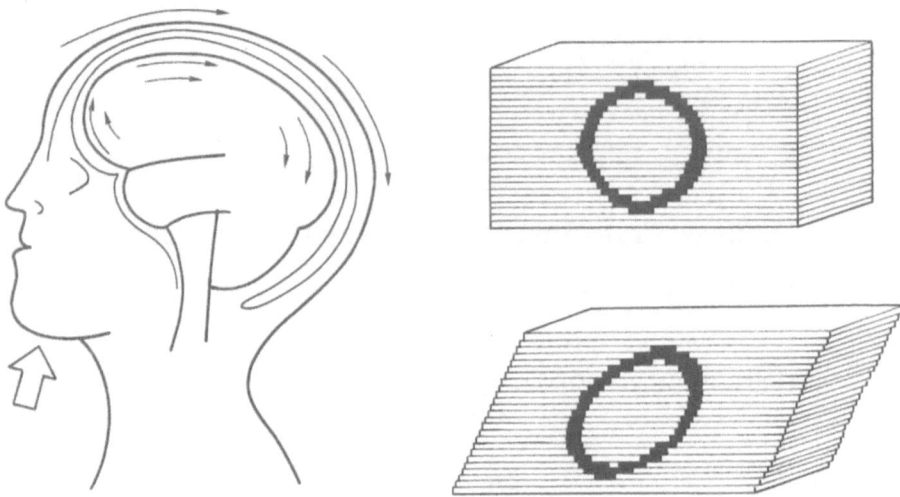

Fig. 1. Shear strain mechanism   Fig. 2. Schematic illustration showing
                                      transformation of the brain by shear strain

the component tissues (Fig. 1). Its intensity is partly a function of distance
from the center of rotation. The brain itself is not compressible but is easily
deformed by shear strain (Fig. 2). Crompton (1971) and Mitchell and
Adams (1973) stated that primary brainstem damage was always accompanied by other cerebral damage.

The neuropathological findings of hemorrhages in the basal ganglia,
subependyma of the ventricles, corpus callosum and/or the brainstem
(Figs. 3, 4 and 5) can be recognized in the CT scan of the surviving brain.
There may be ischemic lesions in the same regions. Diffuse brain swelling is
considered to be another manifestation of diffuse injury (Table 1). Focal
injury consists of localized brain damage resulting from impact. Cortical
contusion is its commonest manifestation (Fig. 6). Subdural or intracerebral hematoma also represent focal injury. Focal and diffuse brain injury
may coexist following impact (Fig. 5).

Experimental Confirmation of Diffuse Brain Injury

Our head injury research group from Tokyo Jikei University as well as
the Woman's Medical College and the Engineering Faculty of Tokyo
University developed a human head model. The skull consists of plastic
material covered by synthetic rubber. The dynamic characteristics of these
materials were almost the same as those of a human skull and scalp, and the

Table 1. *Concept of Focal or Diffuse Brain Injury in Correspondence with Neuro-pathological Descriptions*

Brain injury	Clinical and neuropathological findings
Focal Brain Damage	epidural hematoma subdural hematoma intracerebral hematoma cortical contusion

Diffuse Brain Damage	white matter hemorrhage
	brainstem hemorrhage
	cerebellar lesion
	late sequelae
	   diffuse degeneration of
	   the white matter
	
	concussion
	diffuse brain swelling

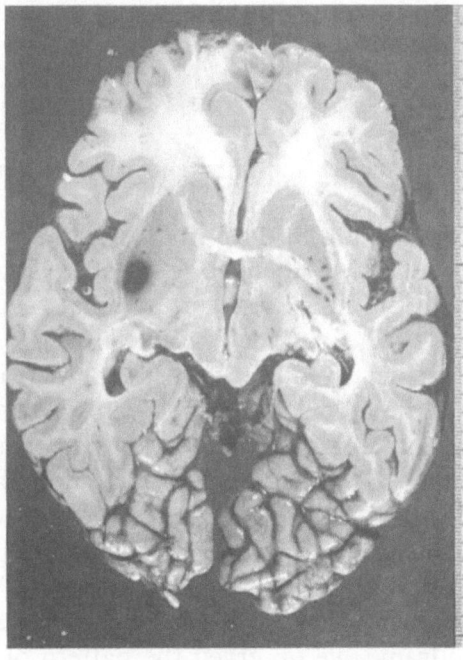

Fig. 3. Basal ganglia hemorrhage

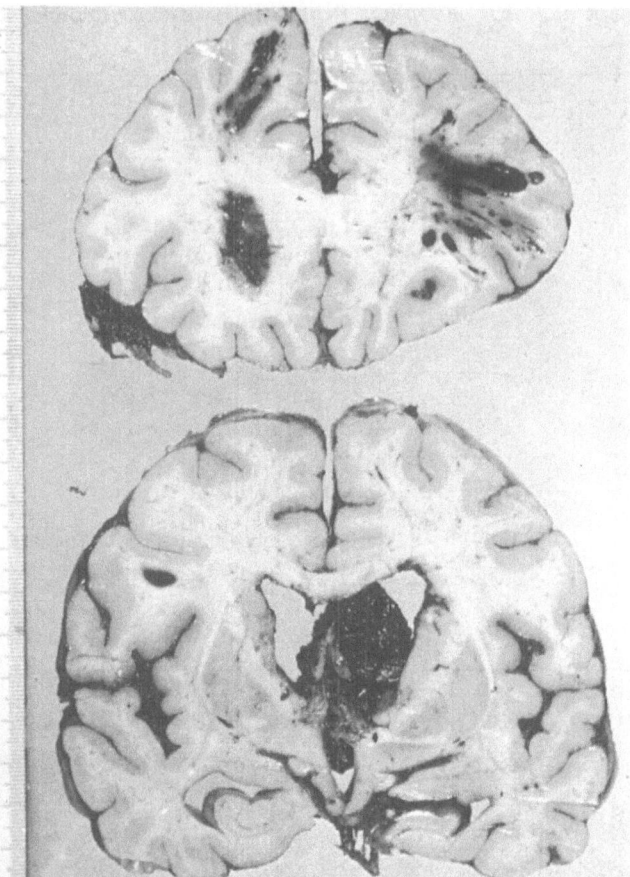

Fig. 4. White matter and intraventricular hemorrhage

model was life-size (Figs. 7 and 8). Two microsensors were fixed inside the skull at the frontal and occipital poles to measure intracranial pressure changes at the moment of impact. The intracranial cavity was filled completely with water; and the head model was fixed to a standard human body dummy. An impact was delivered in increments of increasing magnitude to the midfrontal area of the model (Hayashi 1969). The negative pressure at the occipital pole measured at the moment of impact was almost never less than one atmosphere, which indicated that cavitation had occurred at the antipode (Fig. 9). About 100 g impact was thought to be sufficient to produce cavitation. But it was doubtful if such cavitating formation also occurred in normal brain.

In a second two-dimensional brain model made of gelatine, packed in a transparent plastic frame, we observed the pattern of shearing strain in the gelatine when a *rotational acceleration* impact was delivered to the

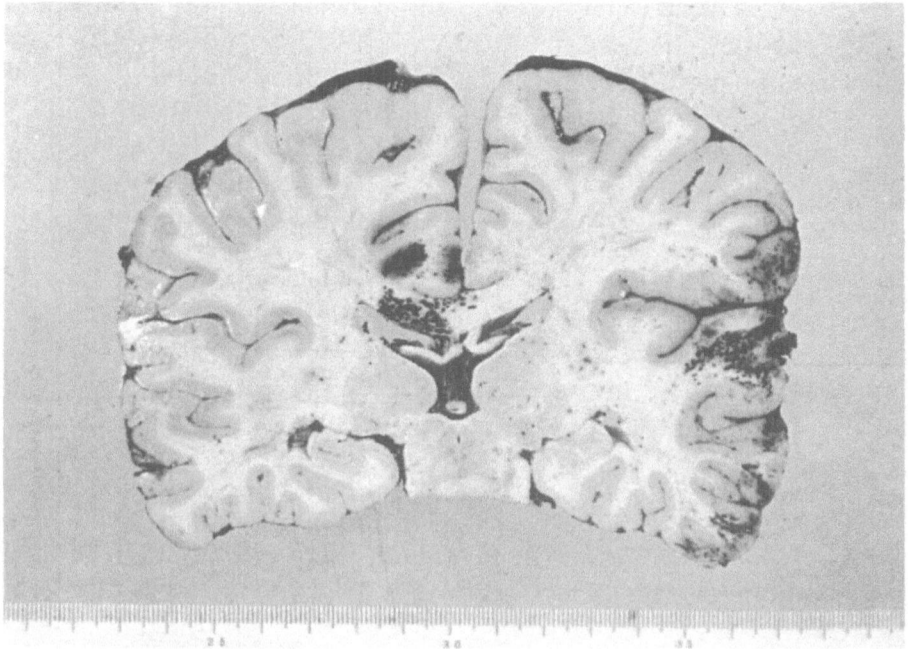

Fig. 5. Corpus callosum and cingulate gyrus hemorrhage in combination with cortical contusion

frame. The shear strain was most marked at the points corresponding to the skull base and the surface of the cerebral convexity (Fig. 10).

In further investigations (Kanda *et al.* 1981, Masuzawa *et al.* 1976, Nakamura *et al.* 1981, 1984, 1986, Ono *et al.* 1980, 1988) an anesthetized monkey was placed on a sledge to apply *translation acceleration* impact to its head (Fig. 11).

1110 g of averaged acceleration, at most, with 4.3 millisecond duration impact was delivered to the head of the monkey, but no focal or diffuse damage or hemorrhage was found in any of the animals by macroscopic and light microscopic examinations. It meant that it was doubtful that a translation acceleration impact would give rise to cavitation phenomena at a site opposite to the impact and also to coup and contre-coup contusions. This observation was contrary to the results obtained from the dynamic experiment using the head model; and it meant that water did not simulate viscoelastic brain tissue in the study of brain trauma.

When *rotational acceleration* impact was delivered to the anesthetized monkey's head in sagittal orientation (Fig. 12) an occipital impact based on averaged accleration of 500 g and a duration of 2.7 millisecond was found to produce a contre-coup contusion of the frontal brain (Fig. 13). *Gliding*

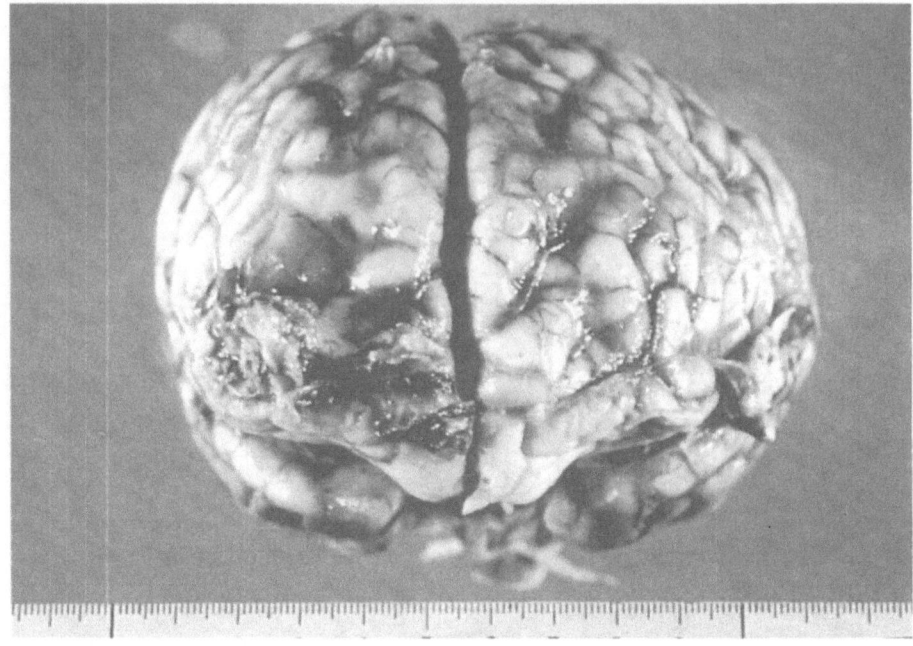

Fig. 6. Cortical contusion

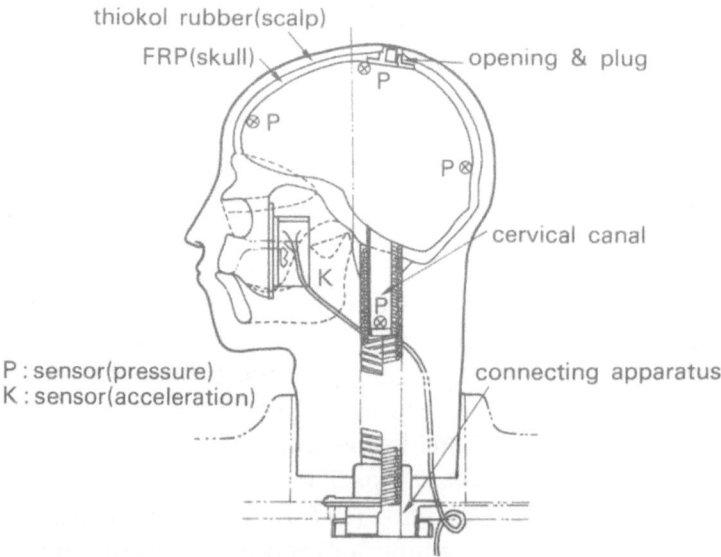

Fig. 7. Human head model developed by author and colleagues

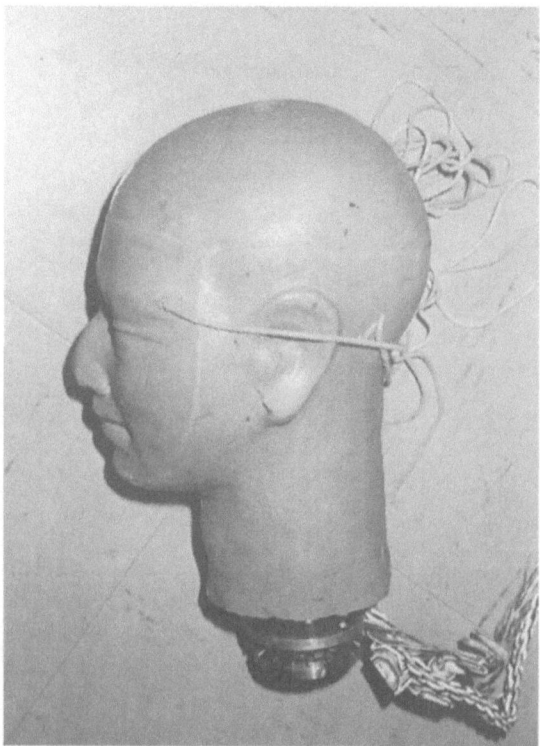

Fig. 8. Photograph of life-size dummy illustrated in Fig. 7

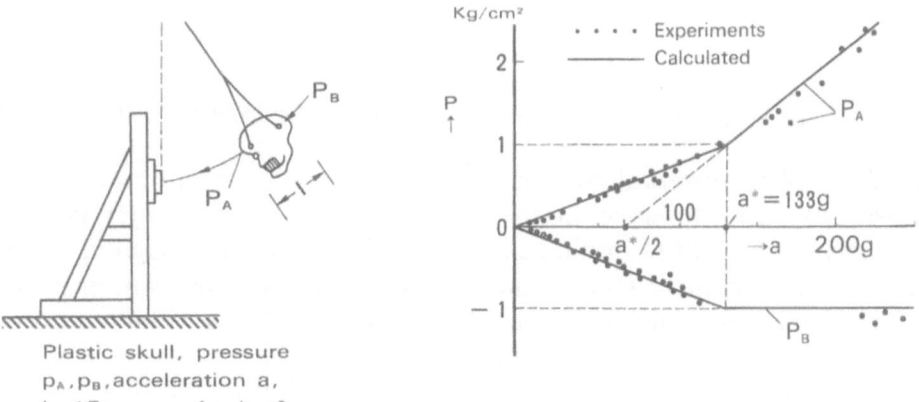

Fig. 9. Results of experiment with human head dummy

N. Nakamura

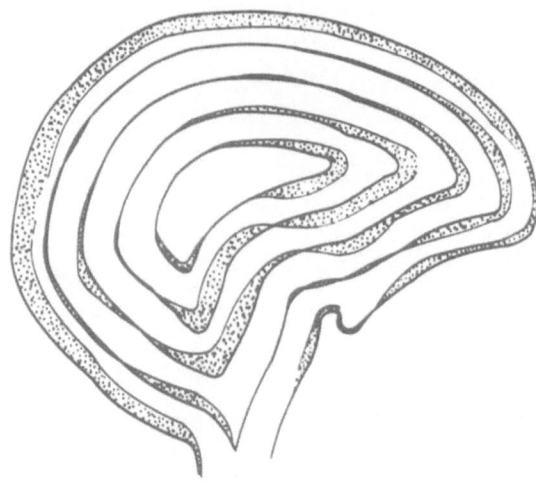

CLOCKWISE ROTATION

Fig. 10. Magnitude and extent of shear strain observed in two-dimensional
gelatin model

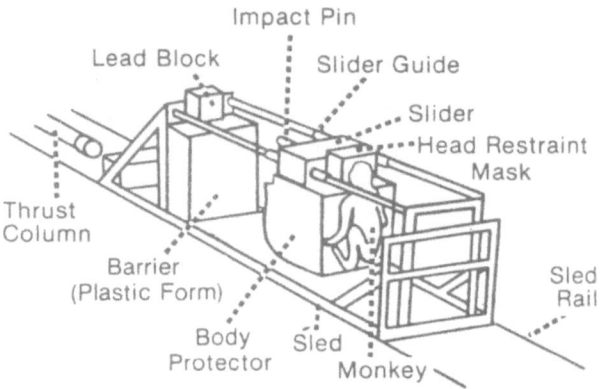

Fig. 11. Experimental apparatus for newly developed impact system for the first part

contusion of the cerebral convexity, reported by Lindenberg and Freitag
(1960), was also observed in the monkeys (Fig. 14). The latter type of
contusion is thought to originate from stretching of the parasagittal bridging veins due to shear strain, and corresponds to our observation in the
gelatin model experiment. But even an impact of 850 g × 4 ms did not
produce an intracerebral hemorrhage.

Rotational impact was delivered to the temporal region of monkey's
head as the third stage of the experiment (Fig. 15). This impact resulted in
both cortical contusions and deep intracerebral hemorrhages. The latter

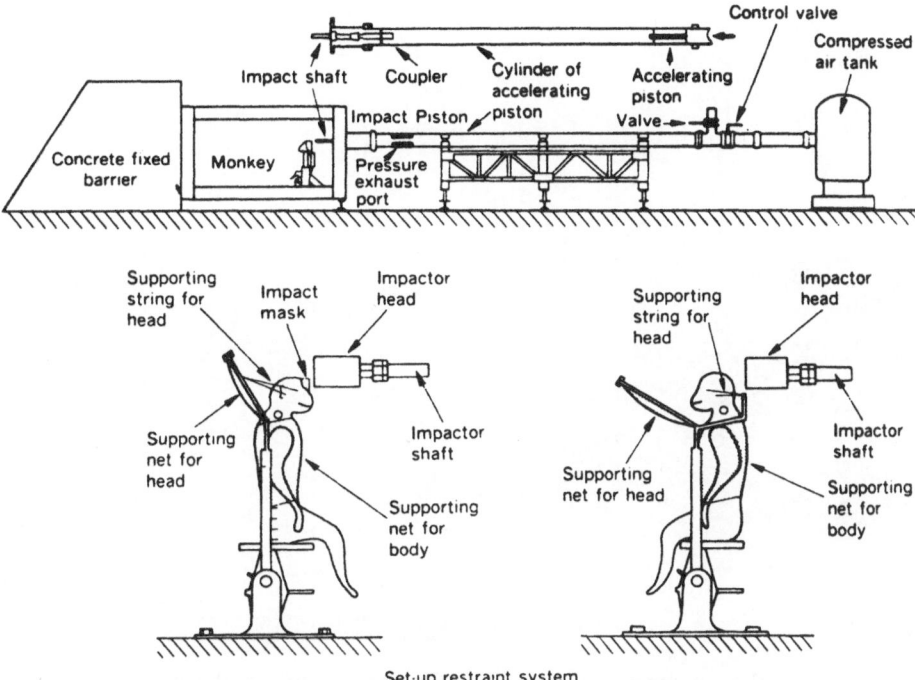

Fig. 12. Experimental apparatus for the second part

included hemorrhages in the corpus callosum, basal ganglia (Fig. 16), subependymal layer (Fig. 17) and brainstem. In some of the animals Evans blue dye was injected intravenously soon after the impact, with staining of the subcortical white matter in blue. This brain damage occurred after impacts of 800 g × 3.2 ms or 550 g × 5 ms, which were smaller than in the first stage of the experiment but a little greater than in the second stage.

Thus we come to the result, that since even the strongest translation acceleration impact did not result in either focal or diffuse brain damage, both cavitation and deforming mechanisms are unlikely to play a primary role in the development of cortical contusions or deep hemorrhages. Although some authors believe that these two mechanisms are implicated in focal brain injury, the results of our experiments did not support this hypothesis.

On the other hand, our experiments suggest that rotational acceleration impaction induces localized, severe shear strain between the basal convexity of the frontotemporal lobe and the uneven skull base which results in cortical contusion. Thus, from a dynamic perspective, cortical contusion may be referred to as a focal shearing injury.

When the rotational acceleration impaction is severe enough to cause

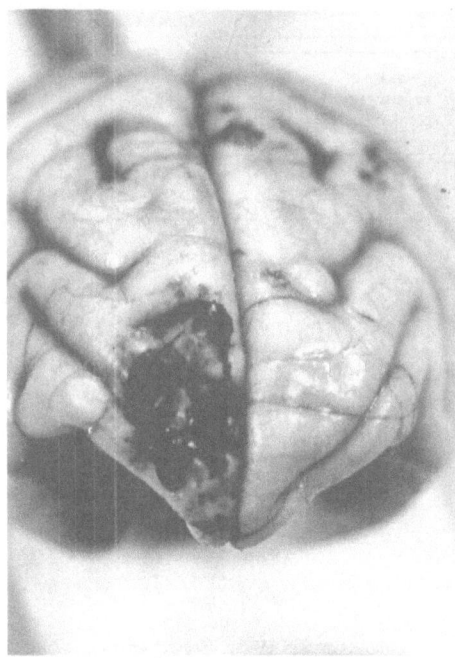 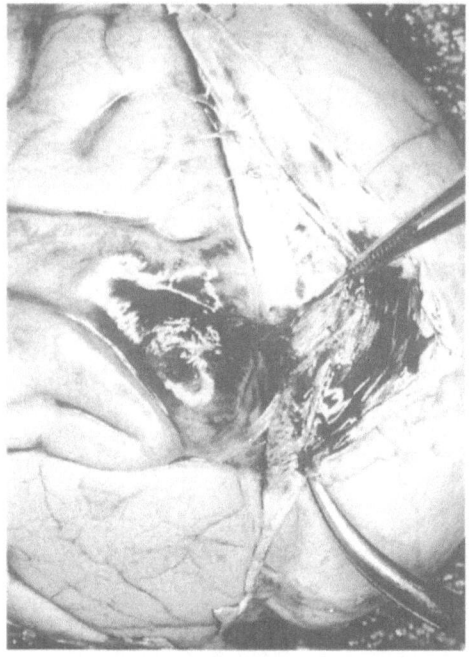

Fig. 13. Cortical contusion (monkey brain)   Fig. 14. Gliding contusion (monkey brain)

a shear strain all over the brain, deep hemorrhages occur. The shear strain must be at its point on the surface of the brain. It also affects deep structures, including the brainstem, and is concentrated on the boundary zone between different structures, as the periventricular zone, corpus callosum and brainstem. Thus diffuse brain injury may be considered a diffuse shearing injury.

Ommaya and Gennarelli (1974) after a study similar to our animal investigations, and examining the SEP as well as the distribution of pathological lesions in the brain, proposed the paradigm given in Fig. 18: Rotational components of accelerative trauma to the head produce a graded centripetal progression of diffusive cortical – subcortical disconnection phenomena – which are always maximal at the periphery and enhanced at the sites of structural changes.

Cerebral Concussion, Diffuse Brain Swelling, Diffuse Degeneration of White Matter and Gliding Contusion as Parts of "Diffuse Brain Injury"

The dynamic mechanism and the pathophysiology of concussion are, to date, not yet clearly known, but the concussion observed in our animal

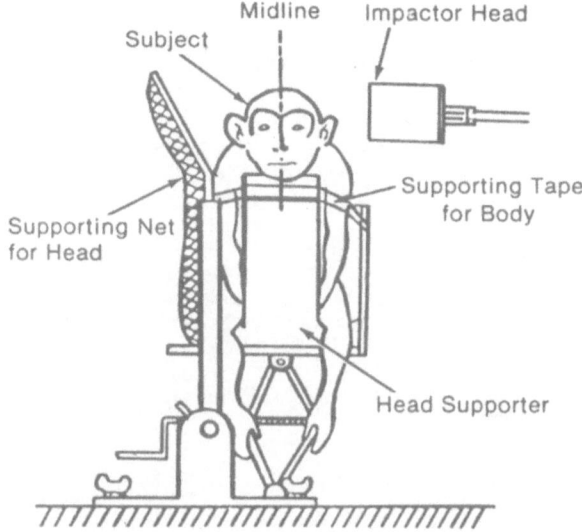

Fig. 15. Experimental apparatus for the third part

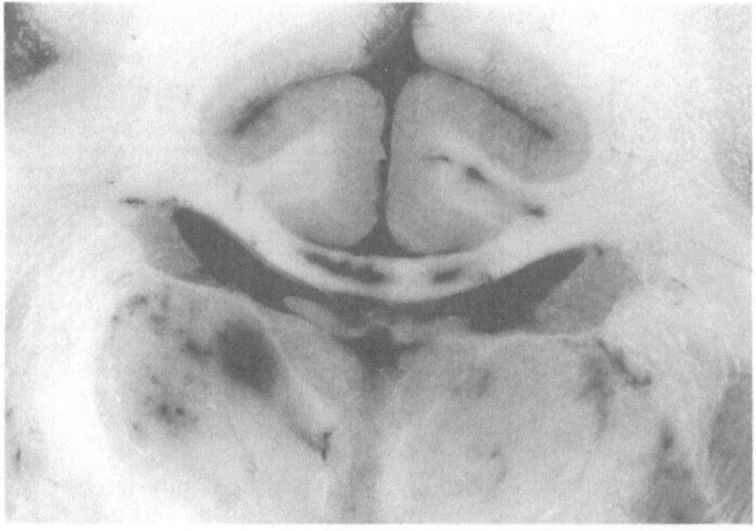

Fig. 16. Corpus callosum and basal ganglia hemorrhage (monkey brain)

experiments (Nakamura *et al.* 1984) demonstrated immediate and temporary dysfunction of the brainstem activity by studies of the corneal reflex, respiration, blood pressure, pulse rate and eye movements. Continuous EEG monitoring during the experiments provided evidence that the impact also affected cerebral activity.

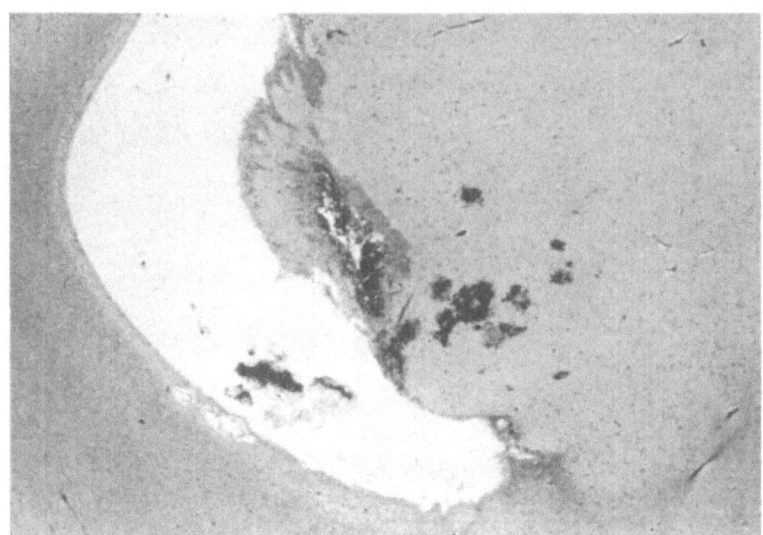

Fig. 17. Subependymal hemorrhage (monkey brain)

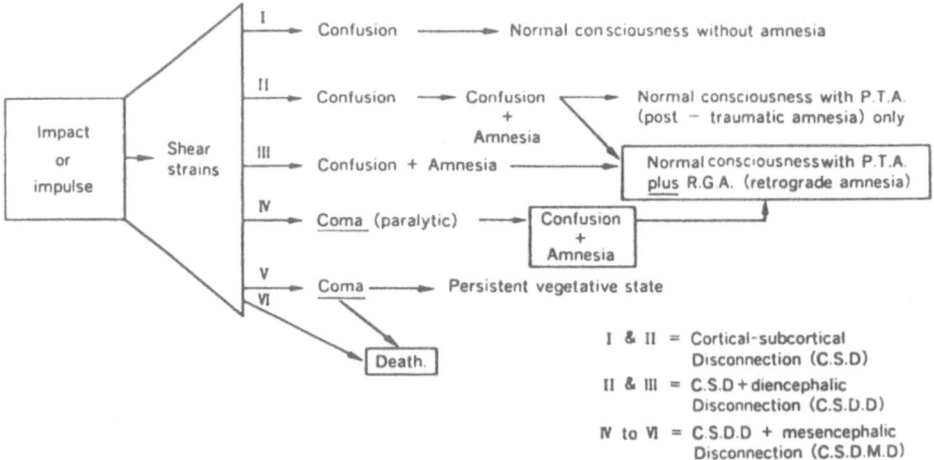

Fig. 18. Schematic paradigm proposed by Ommaya *et al.*

These dysfunctions were aggravated by subsequent systemic circulatory insufficiency, which may prolong loss of consciousness. These symptoms occurred in the absence of microscopic intracerebral hemorrhage or demyelinating lesions. The findings suggest that concussion is the mildest type of diffuse shearing injury.

Our appreciation of *diffuse brain swelling* has changed with the use of the CT. As some fatal head injuries with severe brain edema are not

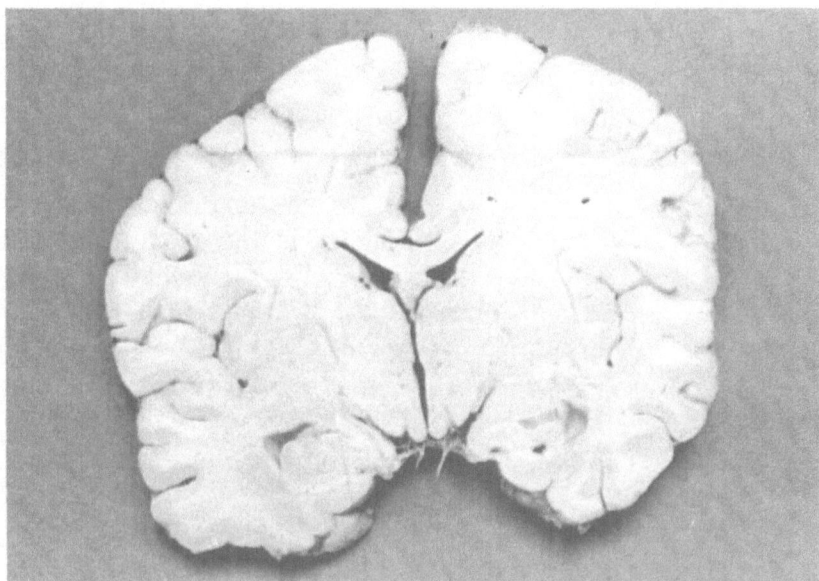

Fig. 19. Macroscopic appearance of diffuse brain swelling

associated with massive intracerebral hemorrhage (Figs. 19 and 20) this condition is regarded as a process of diffuse brain swelling (Bruce *et al.* 1981, Gennarelli 1982, Giannota and Weiss 1981, Sweet *et al.* 1978, Zimmermann *et al.* 1978). Because the brain shows no focal shearing damage and the vasomotor dysfunction is diffuse all over the brain, it may be reasonable to consider diffuse brain swelling to be a type of diffuse shearing injury.

Traumatic *diffuse degeneration* of the white matter (Fig. 21) was presumed by Strich (1956, 1961) to be the end result of widespread disruption of nerve fibers in the white matter occurring at the moment of impact. Adams *et al.* (1983) concluded from experimental and clinical observations, that this condition arises from diffuse axonal injury, which is the most serious type of diffuse injury.

Finally, *gliding contusion* (Figs. 22 and 23) was observed by Lindenberg and Freytag 1960, at the parasagittal convexity. We found it in some monkey brains (Fig. 16) subjected to sagittal rotational impaction. Lindenberg discussed severe shear strain, produced between the dura and the brain by the impact, stretching the parasagittal bridging veins abruptly. As a result, the veins are ruptured and the nearby arachnoid membrane and the cerebrum are torn. Therefore gliding contusion may be regarded as a focal rather than a diffuse shearing injury. To date, this type of contusion has drawn only little attention, perhaps because it results in almost no clinical symptoms unless an intracranial massive hematoma develops.

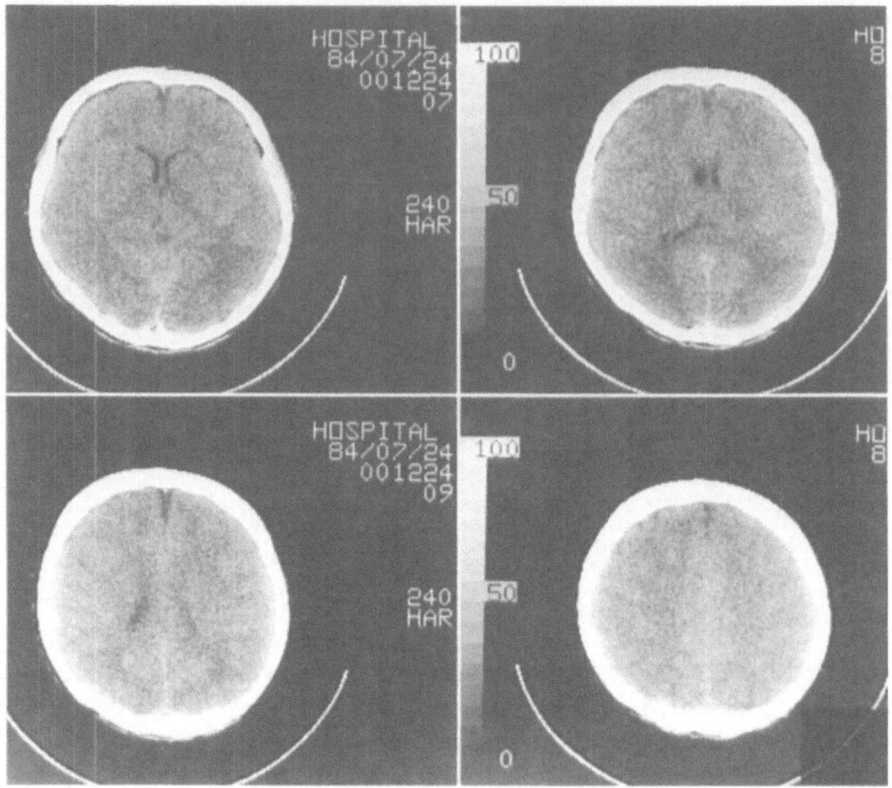

Fig. 20. CT image of diffuse brain swelling

CT-scans do not demonstrate gliding contusions unless the scan extends up to the upper vertex where such contusions are most likely to occur.

The five patients with gliding contusions whom we have encountered had brief posttraumatic loss of consciousness but no deep-seated hemorrhage on the CT. All recovered uneventfully.

Clinical Aspects of Diffuse Brain Injury

Gennarelli (1982) classified diffuse brain injury into five categories:

Mild concussion means that the patient does not lose consciousness but shows temporary neurological dysfunction, including nausea, dizziness, visual loss, motor weakness, etc., immediately following the head injury.

Classical concussion implies that the patient has initial unconsciousness for less than 6 hours and may involve cortical contusion. The patient may have memory loss or sensorimotor complaints.

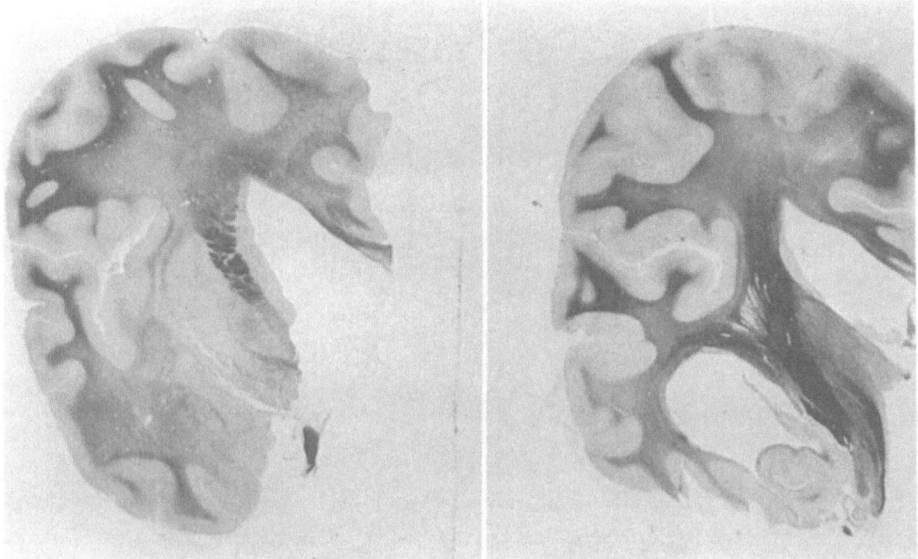

Fig. 21. Diffuse axonal injury or traumatic diffuse degeneration of white matter. The 62-year-old patient survived for 8 months

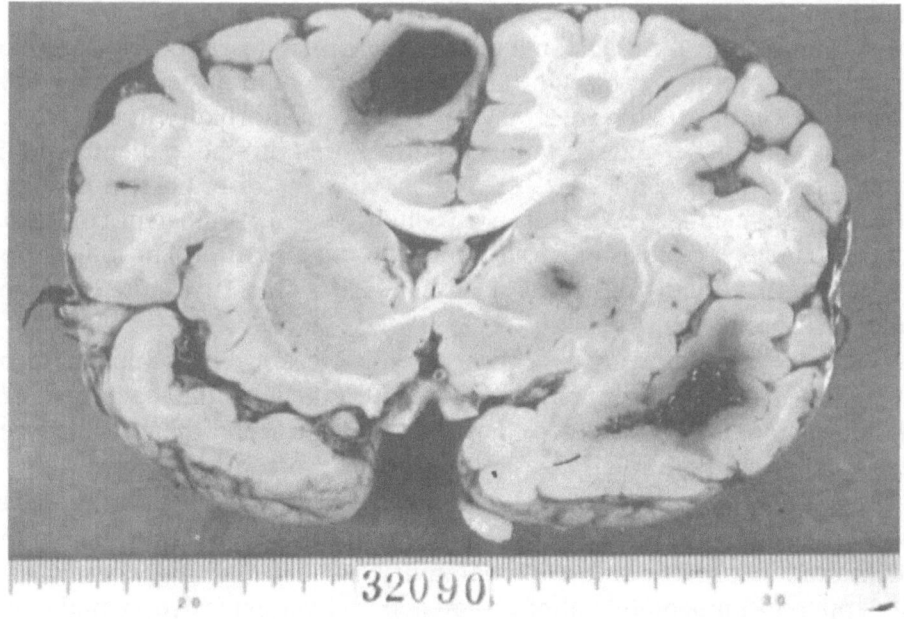

Fig. 22. Gliding contusion consisting of small intracerebral hematomas

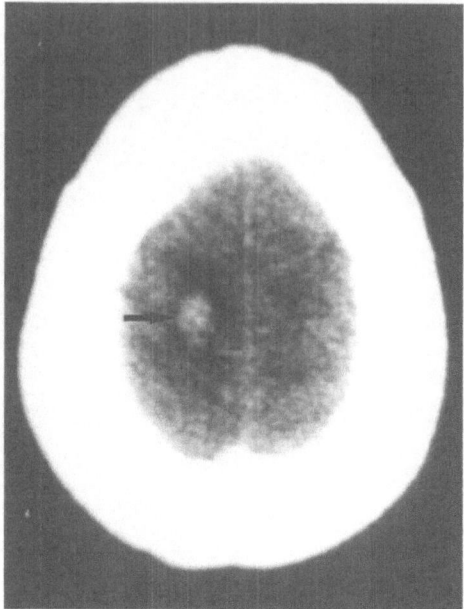

Fig. 23. CT image of gliding contusion

A mild diffuse axonal injury can be defined as a state in which a patient loses consciousness for 6 to 24 hours. It may involve long-lasting or permanent neurological or cognitive deficits.

With moderate diffuse axonal injuries the patient loses consciousness for more than 24 hours and bears little or no evidence of brainstem dysfunction. This results in up to 20% mortality or significant morbidity in survivors.

A severe diffuse axonal injury is diagnosed in the patient who is comatose for more than 24 hours and shows signs of brainstem dysfunction. High fever and hypertension are often present. Brainstem contusion was the clinical diagnosis previously made in this situation. The majority of the patients eventually die, some survive, but in a vegetative state or a poor functional prognosis.

Gennarelli's clinical classification of diffuse brain injury appears to be useful but is not yet in common use. The author believes that each category of diffuse brain injury results from an extensive intracerebral shearing force resulting from a rotational acceleration at the moment of impact.

In a mild and moderate diffuse axonal injury, clinical and pathological evidence demonstrates shearing damages primarily in the cerebrum. On the other hand in a severe diffuse axonal injury, brainstem damages are clinically apparent immediately after an impact, and yet extensive axonal damages can be recognized all over the brain in an autopsy.

Therefore, the third group together with the fourth would be appropriately termed "cerebral shearing injury" and the fourth group alone "diffuse axonal shearing injury".

In spite of only a mild head injury, on CT scans small lesions in the brain are frequently found in the tip and/or base of the frontal and/or temporal lobes adjacent to the skull.

Our group (Sekino, Nakamura et al. 1981) studied 219 consecutive cases of head injury regardless of severity and carried out CT examinations within several days after trauma.

Out of 171 patients who remained alert or lost consciousness for less than 10 minutes, 13 (7.6%) had intracerebral lesions. This figure may be somewhat overestimated, since it reflects only patients who attended hospitals and underwent CT scans. Three out of 12 patients (25%), who were unconscious for 10 to 60 minutes after trauma, showed intracerebral lesions in the CT scan. In every patient who lost consciousness for more than 6 hour after trauma, the CT scan invariably demonstrated cerebral lesions.

A few patients lose consciousness for a long time, unaccompanied by any CT scan abnormality (Lobato et al. 1986; Frowein et al. in this volume).

Regarding those patients with a very severe head injury, Adams et al., 1983, investigated 45 cases of diffuse axonal injury (DAI) and 132 cases of non-diffuse axonal injuries (non-DAI) (Table 2). The pathology of the DAI was characterized by focal lesions in the corpus callosum and in the dorsolateral quadrant of the rostral brainstem adjacent to the superior cerebellar peduncles and with evidence of diffuse damage to axons. The predominant features included absence of a clinically manifest interval of posttraumatic lucidity. Skull fracture and intracerebral hematoma were less

Table 2. *Diffuse Axonal Injury (DAI) and Non-DAI in Autopsy Findings Reported by Adams et al. 1983*

		Diffuse axonal injury (DAI) 45 cases	Non-DAI 132 cases
Lucid interval		(−)	(+) or (−)
Diffuse brain swelling	unilateral	1	17
	bilateral	6	7
Fracture		13	114
Intracranial hematoma		5	88
Intracranial hypertension		25	114
Hypoxic damage	multifocal	0	10
	boundary	18	25

frequent in DAI than in non-DAI. Clinical and pathological evidence of intracranial hypertension was less common in DAI and intracerebral is-chemia lesions were numerous in non-DAI. In DAI brain swelling was diffuse and bilateral, whereas it was unilateral in non-DAI.

Thus, we propose the following classification of diffuse brain injury: 1) cerebral concussion, 2) cerebral shearing injury, 3) diffuse axonal (shear-ing) injury.

Diffuse brain swelling is also included as diffuse brain injury. However, it can be diagnosed only by CT scan.

Diagnostic Criteria

1. Cerebral Concussion

The International Neurotraumatology Committee defined concussion as a clinical syndrome characterized by immediate impairment of neural functions such as alterations of consciousness, disturbance of vision, motion and sensation due to mechanical forces (Gurdjian *et al.* 1979). Although the clinical symptoms are usually transient, in more severe concussion the symptoms may be severe and persist for varying length of time.

When a patient is initially unconscious for 6 hours or longer after trauma, evidence of an intracerebral hemorrhage will most likely be shown on the CT.

Therefore, the author proposes that a period of unconsciousness of less than 6 hours is appropriate.

The term "cerebral concussion" is suitable for clinical use whether the patient has a small intracerebral hemorrhage or not.

The transient alterations in neurological and vital signs and mental disorders are described in the chapter by Vigouroux in this volume.

2. Cerebral Shearing Injury

Apart from cerebral contusion, which I consider a focal shearing injury, the cerebral shearing injury is a moderate form of diffuse brain injury.

The patient will be unconscious, for 6 hours or more, without a lucid interval.

In a 58-year-old patient the exemplary CT scan showed small hemor-rhages only at the left temporal and right frontal lobes (Figs. 24 and 25). However, the unconsciousness lasted 2 days.

With cerebral shearing injuries patients usually recover consciousness within 3 weeks at the most. Final recovery depends on the extent and

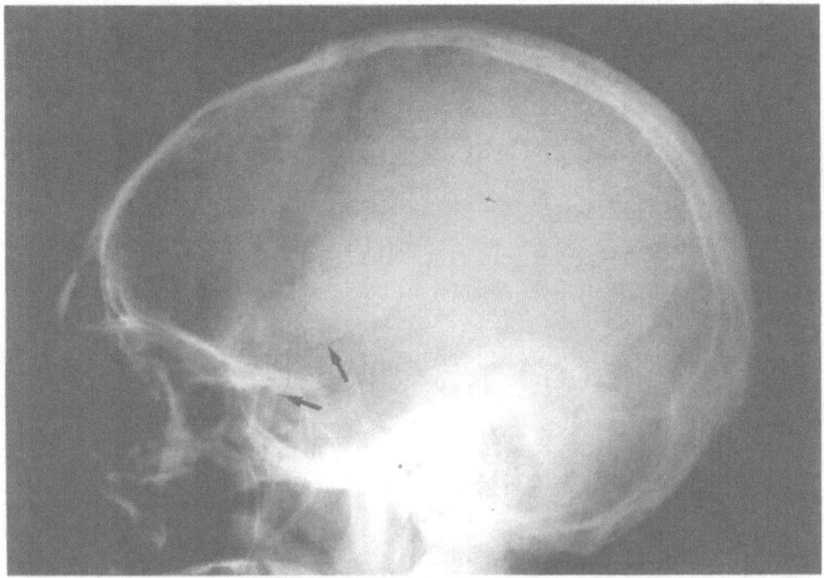

Fig. 24. Temporal skull fracture

location of anatomical damage, however, secondary complications, such as thrombosis of intracranial veins, may affect the outcome.

3. Diffuse Axonal Shearing Injury

This term corresponds to the diffuse white matter shearing injury proposed by Gennarelli, and it is the most severe form of diffuse brain injury.

An example is a 33-year-old jockey, who was thrown off his horse and struck his forehead and face on to the ground. He remained comatose and decerebrate for 16 days and was then in a vegetative state for 8 months until his death. The macroscopic autopsy findings are shown in Fig. 26. The brain weight was 1280 g. There was an old left occipital subarachoidal hemorrhage and an "Etat vermoulu" at the surface of the left parasagittal lobe. All the ventricles were markedly dilated symmetrically. Small softenings or cysts were seen along the sagittal plane between the left caudate nucleus and the internal capsule, in the left temporal lobe, left putamen, inferior part of the left lenticular nucleus, right pre- and post-central gyri, right insula and parahippocampal gyri bilaterally. All these lesions were surrounded by hemosideric tissue. The basal ganglia and thalamus were

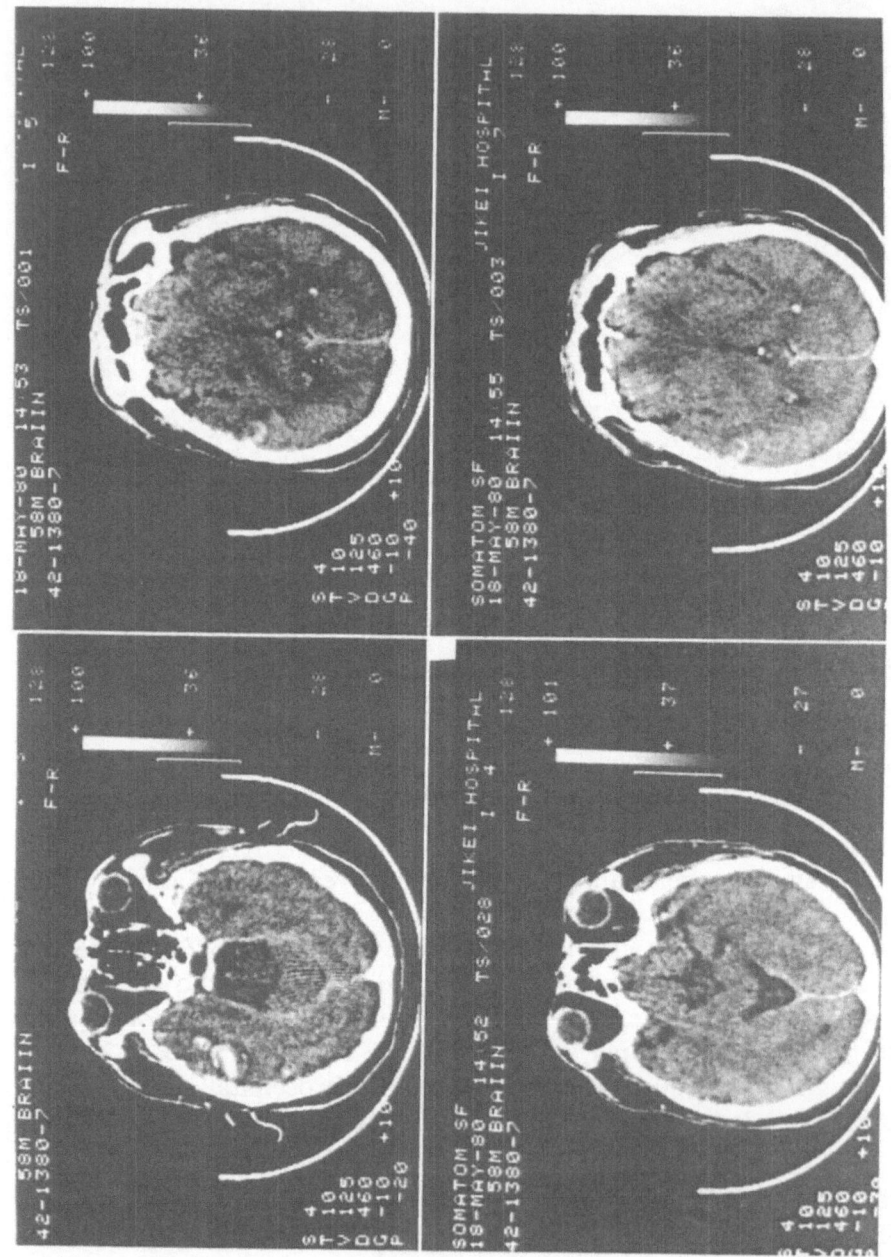

Fig. 25. Cerebral shearing injury associated with small contusion

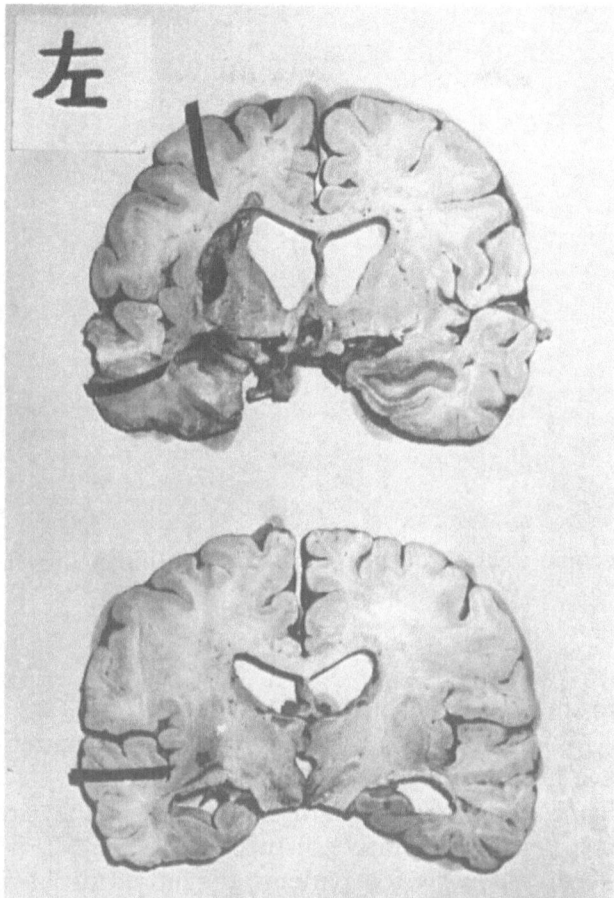

Fig. 26. Diffuse axonal shearing injury

markedly atrophied. There was evidence of an old transtentorial herniation. The cerebellum and brainstem were not characteristically altered.

Microscopic findings: The dilated perivascular space and the white matter showed "status spongiosus and lacunaris" predominantly at the base of both of the left frontal and temporal lobes. The gliosis was extensive in the whole brain, and granular cells were abundant in the white matter. The germistocytic astroglia were characteristic in the left frontal lobe, and some were binucleate. The oligodendroglia were atrophic. The softened areas and cysts showed microscopic evidence of an old hemorrhage accompanied by hemosiderosis and accumulation of granular fat cells. Microscopic small softenings and cysts were scattered in some places.

Thus, the strong impact transmitted directly to the whole brain resulted in extensive degenerations of axons, demyelination within white matter,

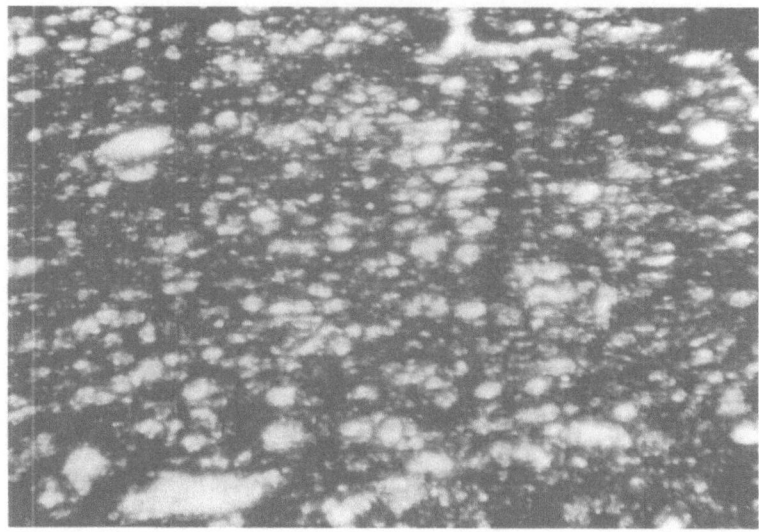

Fig. 27. Microscopic view of diffuse axonal shearing injury (myelin sheath stain of Sugano)

and damage to the cortex, basal ganglia and thalamus (Figs. 27, 28, 29) as described by Ono *et al.* 1983, and many others. The parietal lesion was believed to have been produced by a gliding contusion.

These patients are rendered comatose immediately, they are decerebrate and usually have serious vegetative disturbances.

The long-term *prognosis* for patients diagnosed as having a diffuse axonal shearing injury is, as a rule, very poor. In Gennarelli's (1982) experience the outcome at one month after trauma was a mortality of 55%, vegetative state 36%, and severe deficits 9%.

Among our 10 patients there were two fatal courses, five patients were in a vegetative state and three were severely disabled at 6 month post trauma.

We have no experience with measuring *intracranial pressure* in the early stage. Adams (1983) reported that the incidence of raised intracranial pressure was 56% in cases with diffuse axonal injury, as shown by the presence of pressure necrosis in one or both parahippocampal gyri.

In my experience, at one month or more after trauma the intra-cranial pressure was at the lower limit of normal or much lower, which suggests CSF-hypotension due to cerebrocirculatory insufficiency.

CT scans performed soon after trauma clearly demonstrate the presence of hemorrhages in the brainstem (Figs. 30–32).

Recording the *electroencephalogram*: serial EEG examinations in the same patient sometimes reveal good recovery in long surviving patients,

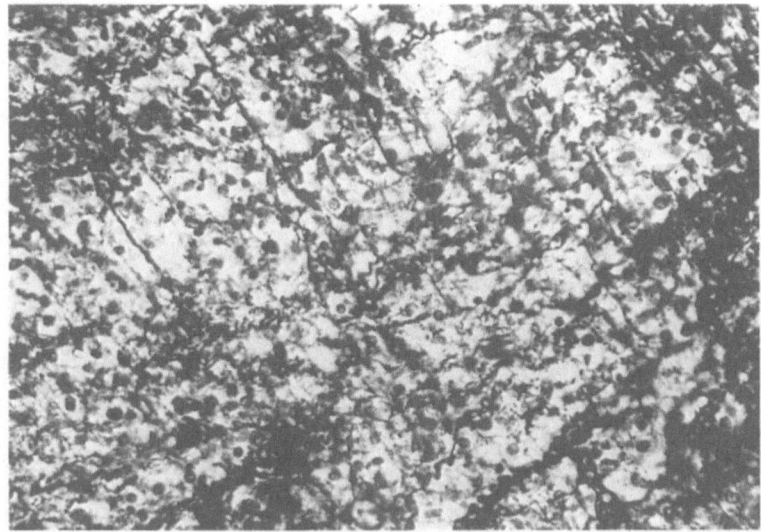

Fig. 28. Microscopic view of axonal degeneration (Bielschowsky's stain)

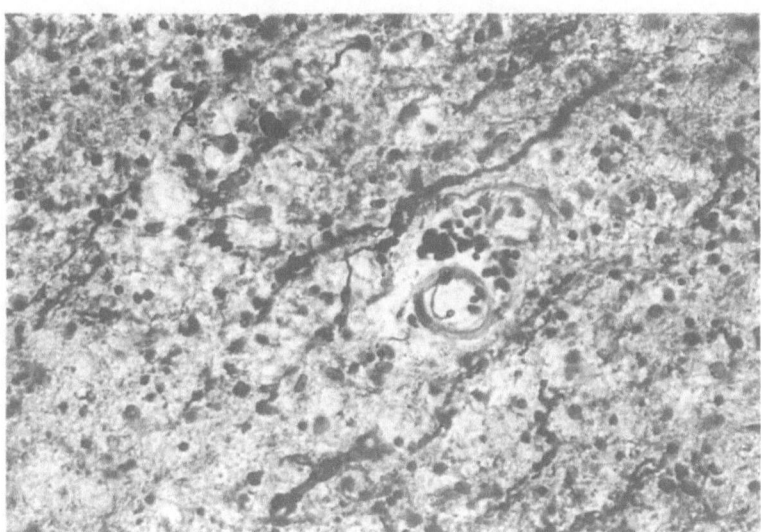

Fig. 29. Microscopic view of axonal degeneration (Bielschowsky's stain)

with delayed responses becoming successively less marked (Fig. 33). Clinical recovery, however, does not always correspond to the EEG recovery.

Far field *brainstem auditory evoked responses* (BAEP) are said to be a reliable tool in the diagnosis of diffuse axonal shearing injury, as well as a good predictor of outcome (see the chapter of Firsching in this volume).

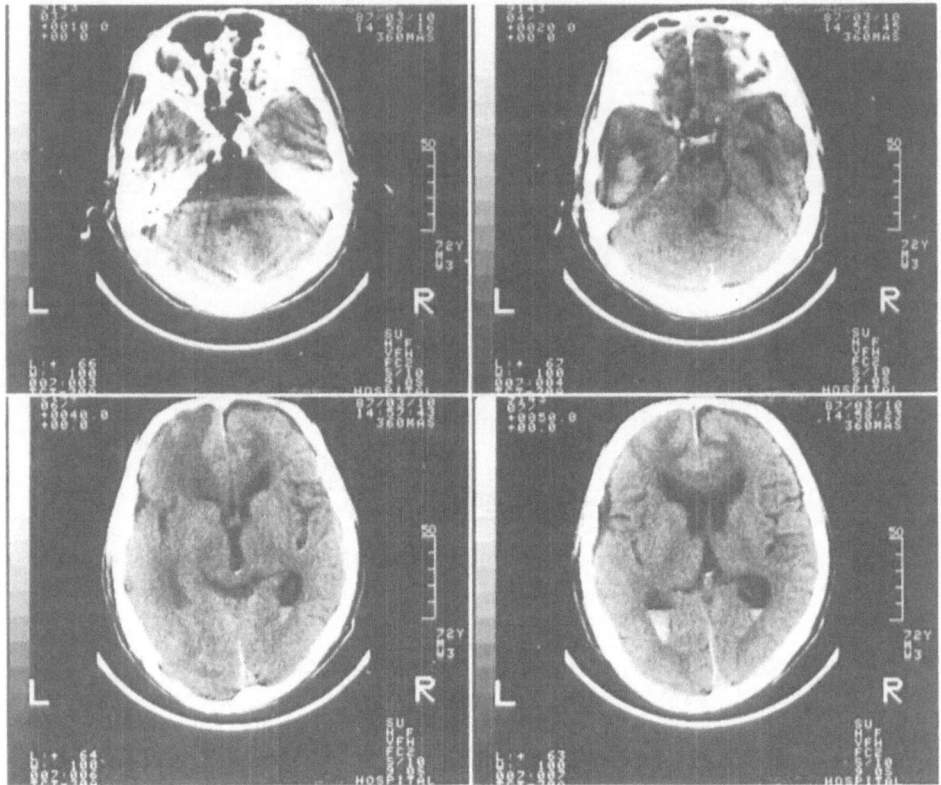

Fig. 30. Multiple intracerebral and intraventricular hemorrhages in diffuse axonal shearing injury

Wave V latency, which reflects activity at the inferior colliculus, is significantly delayed in patients with a shearing injury as compared to other groups.

Kobayashi (1984) investigated 11 patients diagnosed clinically as having a primary brainstem injury. The absence of waves II–V was predictive of death. Some patients, however, showed a normal BAEP at the first examination.

Primary Brainstem Injury – Brainstem Contusion

On the basis of autopsy studies, Mitchell and Adams 1973, found that primary brainstem damage was always associated with other traumatic lesions.

Zuccarello *et al.* (1983) investigated 36 patients with primary brainstem hemorrhage visualized by CT scan and corresponding closely with the

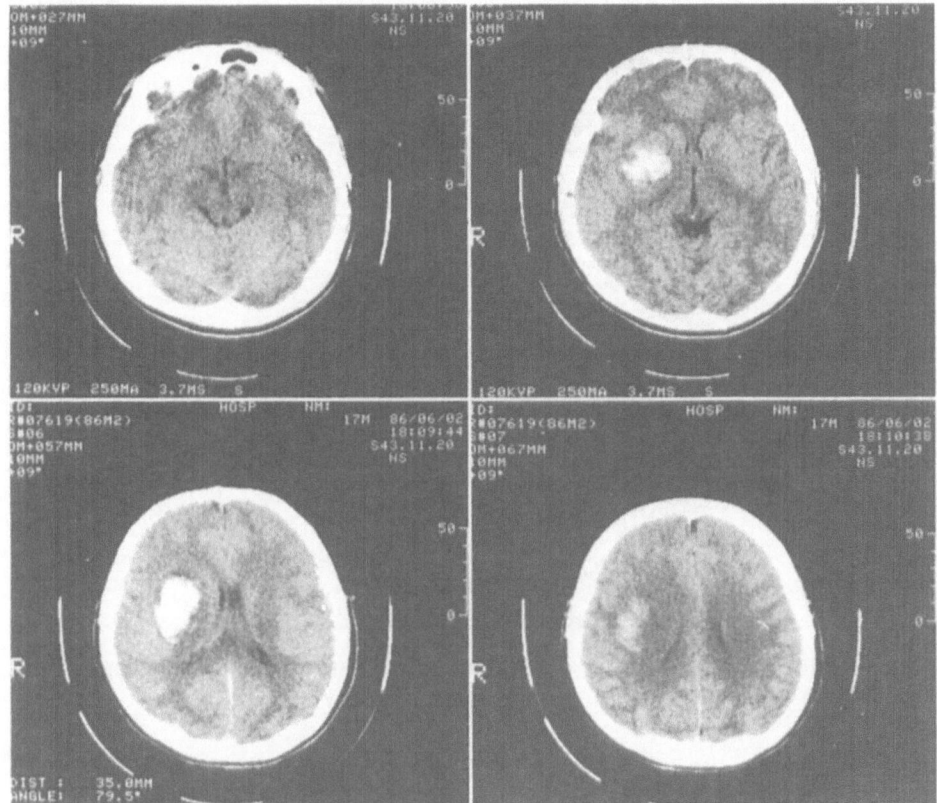

Fig. 31. Basal ganglia hemorrhage in diffuse axonal shearing injury

clinical course. The autopsies of 25 of these patients confirmed brainstem hemorrhage and also disclosed that it was always associated with multiple hemorrhages in the deep white or gray matter and in the corpus callosum.

Cooper *et al.* (1979), reporting 7 cases of traumatic brainstem hemorrhage, noted that when the first CT scan, obtained within 24 hours, showed brainstem hemorrhage, there were invariably multiple intracerebral hemorrhages present as well. Tsai *et al.* (1980) discussed 67 cases of clinically diagnosed brainstem injury. CT scans of these patients showed hemorrhage in the brainstem in 12 cases, low density in 6, mixed density in 11, contrast enhancement in 11 and normal images in 27. Tsai *et al.* (1980) stated that brainstem lesions might occur alone or in association with other cranial injuries.

It is reasonable, then, to assume that primary brainstem damage severe enough to result in hemorrhage, is indicative of severe, diffuse brain injury involving multiple intracerebral lesions.

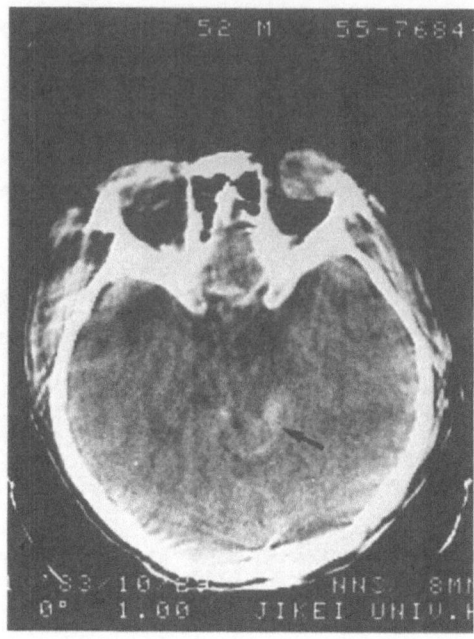

Fig. 32. Brainstem hemorrhage associated with subarachnoid hemorrhage in diffuse axonal shearing injury

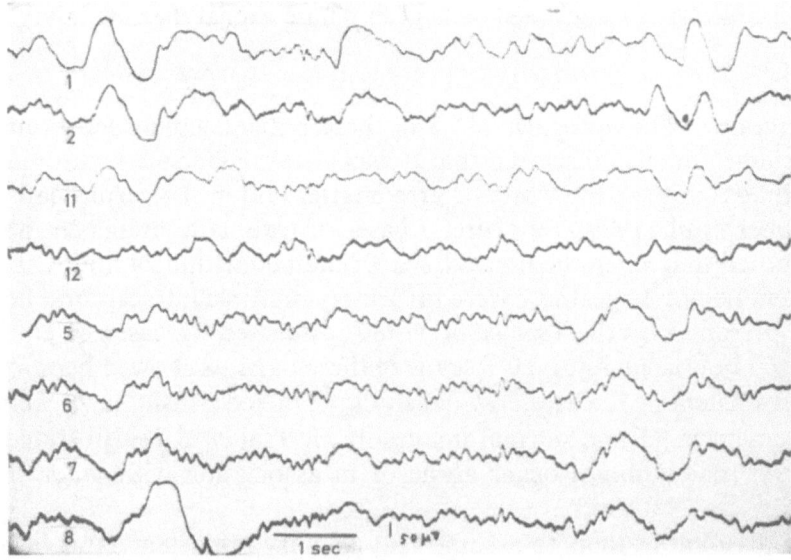

Fig. 33. Electroencephalogram in diffuse axonal shearing injury (5 months post-traumatic)

However, there are rare exceptions in which clinical symptoms are moderately severe or less so, but the CT scan shows primary small brainstem hemorrhage without any other intracerebral abnormalities (Mayer 1967, Lobato *et al.* 1986). It is not clear whether these cases represent a subtype of diffuse brain injury or there is another pathomechanism that causes isolated brainstem hemorrhage.

Finally there are two syndromes peculiar to brainstem contusion: The *medial longitudinal fasciculus syndrome* (MLF) is believed to be a consequence of trauma-induced vascular damage to the midbrain or pons (Sato *et al.* 1974).

With the *acute pontine syndrome,* known as the locked-in syndrome, the patient even though decerebrate, remains conscious, and with intensive treatment these cases are not always hopeless. It rarely arises from trauma. Britt *et al.* (1977) and Turazzi and Bricolo (1977) reported separately that autopsy revealed laceration between the upper cerebellar peduncle and the tegmentum.

References

1. Adams JH, Mitchell DE, Graham DI, Doyle D (1977) Diffuse brain damage of immediate impact type. Its relationship to primary brainstem damage in head injury. Brain 100: 489–502
2. Adams JH, Graham DI, Murry IS, Scott G (1983) Diffuse axonal injury due to nonmissile head injury in humans: An analysis of 45 cases. Ann Neurol 12: 557–563
3. Adams JH, Graham DI, Gennarelli TA (1983) Head injury in man and experimental animals: neuropathology. Acta Neurochir (Wien) [Suppl] 32: 15–30
4. Britt RH, Herrick MK, Hamilton RD (1977) Traumatic locked-in syndrome. Ann Neurol 1: 590–592
5. Bruce DA, Schut L, Bruno LA, Wood JH, Sutton LN (1978) Outcome following severe head injuries in children. J Neurosurg 48: 679–688
6. Bruce DA, Alavi A, Bilaniuk L, Dolinskas C, Obrist W, Uzzell B (1981) Diffuse cerebral swelling following head injuries in children: the syndrome of "malignant brain edema". J Neurosurg 54: 170–178
7. Cooper PR, Maravilla K, Kirkpatrick J, Moody SF, Skalar FH, Diehl J, Clark K (1979) Traumatically induced brainstem hemorrhage and the computed tomographic scan. Neurosurg 4: 115–124
8. Crompton MR (1971) Brainstem lesions due to closed head injury. Lancet I: 669–673
9. Firsching R, Frowein RA Evoked potentials in head injuries. In this volume
10. Frowein RA, Stammler U, Firsching R, Friedmann G, Thun F Early dynamic evolution of cerebral contusions and lacerations – Clinical and radiological findings. In this volume

11. Gennarelli RA (1982) Cerebral concussion and diffuse brain injuries In: Cooper PR (ed) Head injury. Baltimore, Williams and Wilkins, pp 83–97
12. Giannota SL, Weiss MH (1981) Pitfalls in the diagnosis of head injury. Clin Neurosurg 29: 188–299
13. Gurdjian ES, Brihaye J, Christensen JC, Frowein RA, Lindgren S, Luyendijk W, Norlén G, Ommaya AK, Oprescu J, de Vasconcellos Marques A, Vigouroux RP (1979) Glossary of neurotraumatology. Acta Neurochir (Wien), [Suppl] 25. Springer, Wien New York
14. Hayashi T (1969) Study of intracranial pressure caused by head impact. J Faculty of Engineering Univ Tokyo (B) 30: 59–72
15. Holbourn AHS (1943) Mechanism of head injuries. Lancet II: 438–441
16. Kanda R, Nakamura N, Sekino S, Yuki K, Satoh J, Kikuchi K, Sanada S (1981) Experimental head injury in monkeys. Concussion and its tolerance level. Neurol Med-Chir 21: 645–656
17. Kobayashi MS, Yokota H, Kitamura T, Imaya H, Yajima K, Nakazawa S, Yano M, Otsuka T (1984) Diffuse axonal injury. Neurotraumatology 7: 243–248 (in Japanese with English abstract)
18. Lindenberg R, Freitag E (1960) The mechanism of cerebral contusions. A pathologic–anatomic study. Arch Pathol 69: 440–469
19. Lobato RD, Sarabia R, Rivas JJ, Cordobes F, Castro S, Munos MJ, Cavrera A, Barcena A, Limas E (1986) Normal computerized tomography scan in severe head injury. J Neurosurg 65: 784–789
20. Masuzawa H, Nakamura N, Hirakawa K, Sekino H, Matsuno M, Sano K (1976) Experimental head injury and concussion in monkey using pure linear acceleration impact. Neurol Med-Chir 16; Part 1: 77–90
21. Mayer ETh (1967) Zentrale Hirnschäden nach Einwirkung stumpfer Gewalt auf den Schädel. Arch Psychiatr Nervenkr 210: 238–262
22. Mitchell DE, Adams JH (1973) Primary focal impact damage to the brainstem in blunt head injuries. Does it exist? Lancet II: 215–218
23. Nakamura N, Nasuzawa H, Sekino H, Kondo H, Kikuchi A, Ono K (1981) Which is the more severe impact on the head, sagittal or lateral? In: Ommaya AK et al (eds) Head and neck injury criteria. National Highway Traffic Safety Administration USA, Washington DC, pp 61–69
24. Nakamura N, Sekino H, Kanda R, Masuzawa H, Mii K, Ariga R, Kono H, Sugiura M (1984) Dynamic mechanism of concussion in experimental head injury. Neurotraumatology 7: 1–8 (in Japanese with English abstract)
25. Nakamura N (1986) Experimental head injuries due to direct impact acceleration – Head tolerance limit to impact. In: Anthony Sances jr et al. (eds) Mechanisms of head and spine trauma. Aloray Pub, New York, pp 152–235
26. Ommaya AK, Gennarelli TA (1974) Cerebral concussion and traumatic unconsciousness. Brain 97: 633–654
27. Ono J, Yamaura A, Isobe K, Kotaki M, Watanabe Y, Saeki N, Nakamura T, Makino H (1983) Analysis of CT findings in severe head injuries with special reference to hemorrhagic lesions in the corpus callosum, basal ganglion and brain stem. Neurotraumatology 6: 111–226 (in Japanese with English abstract)

28. Ono K, Kikuchi A, Nakamura M, Kobayashi H, Nakamura N (1980) Human head intolerance to sagittal impact. Proc Twenty-Fourth STAPP Car Crash Conf Society of Automotive Engineers, New York, pp 101–160
29. Ono K, Kikuchi A, Kobayashi H, Nakamura N (1985) Human head tolerance to sagittal and lateral impacts. – Estimation deduced from experimental head injury using subhuman primates and human cadaver skulls. In: Head injury prevention – past and present research. Wayne State University, Detroit, pp 35–72
30. Sato O, Sugita K, Ookoshi Y, Suzuki M, Umei M (1974) Lesion of the MLF syndrome following head injury. Neurochirurgia (Stuttgart) 17: 141–145
31. Sekino H, Nakamura N, Yuki K, Satoh J, Kikuchi K, Sanada S (1981) Brain lesions detected by CT scans in cases of minor head injuries. Neurol Med-Chir 21: 677–683
32. Strich SJ (1956) Diffuse degeneration of the cerebral white matter in severe dementia following head injury. J Neurol Neurosurg Psychiatry 19: 163–185
33. Strich SJ (1961) Shearing of nerve fibers as a cause of brain damage due to head injury. Lancet II: 443–448
34. Sweet RC, Miller JD, Liper M, Kishore PR, Becker DP (1978) Significance of bilateral abnormalities on the CT scan in patients with severe head injury. Neurosurg 3: 16–21
35. Tsai FY, Teal JE, Quinn MM, Itabashi HH, Huprich JE, Ahmadi J, Segall HD (1980) CT of brainstem injury. AJNR 1: 23–29
36. Turazzi D, Bricolo A (1977) Acute pontine syndrome following head injury. Lancet II: 62–64
37. Vigouroux RP, Guillermain P, Rabehanta P, Cerebral contusions – lacerations – clinical study. In this volume
38. Zimmermann RA, Bilaniuk LT, Gennarelli T (1978) Computed tomography of shearing injuries of the cerebral white matter. Radiology 127: 393–396
39. Zuccarello M, Fiore DL, Trincia G, DeCaro R, Pardatscher K, Andrioli GC (1983) Traumatic primary brain stem hemorrhage. Acta Neurochir (Wien) 67: 103–113

Cerebral Contusions and Lacerations
A Clinical Study

R.P. Vigouroux and P. Guillermain
with the collaboration of P. Rabehanta

Marseille (France)

Contents

Introduction

Before presenting this study, a few preliminary remarks should be made. When reference is made to the literature, it appears difficult to give a precise clinical definition of traumatic cerebral lesions, and particularly cerebral contusions–lacerations. The classical distinctions between cerebral commotion, contusion and compression has been modified by numerous authors. Some authors take into account the gravity of the clinical state (Bues and Stewart 1947), or the evolution with time (Fasano 1973; Gruner *et al.* 1965), or the mechanism of production of the lesions (Gurdjian *et al.* 1958).

In everyday practice, this distinction no longer has a precise significa-tion: when the notion of commotion is based on a functional lesion of the brain tissue, autopsies and now the CT scan reveal the possibility of macroscopic lesions. When the notion of contusion is based on an anatom-ical lesion of the nervous system, the resulting clinical manifestations depend on the location of the lesion. These manifestations can therefore be "silent" if the lesion affects a "dumb area". Moreover, systematic use of the CT scan can demonstrate the presence of relatively large lesions in patients in a satisfactory clinical state, but with cerebral contusions–lacerations and associated tissue destruction, oedema and hemorrhage to different degrees. They are thus described under various headings in the literature (Botterel and Stewart 1947; Casella *et al.* 1967; Columella 1973; Feld *et al.* 1955; Geuna *et al.* 1967; Guillermain 1970; Obrador *et al.* 1968; Rusu 1972) and one particular aspect or another is considered by the authors. Comparison of different series is consequently difficult especially as acute subdural hematomas, which are frequently associated with these lesions, are also included with them (Fasano 1973; Lazorthes 1973), or the lesions them-selves are integrated into the group of traumatic intracerebral hematomas (Stender and Schulze 1966; Teasdale and Galbraith 1981; Tönnis *et al.* 1965). Therefore, having recently re-examined this question using 833 personal observations selected in relation to precise criteria (angiographic, tomodensitometric and/or surgical), we have included large parts from this study (Vigouroux and Guillermain 1981). Our data generally agree with those found in the literature which cannot be exhaustive, bearing in mind the numerous publications pertaining to lesions of this type (Casella *et al.* 1967; Cohadon *et al.* 1973; Columella 1973; Da Pian *et al.* 1967; Geuna *et al.* 1967; Liguori and Troisi 1966; Stender and Schulze 1966; Tönnis *et al.* 1965).

Symptomatology

Traumatic cerebral lesions are usually shown by alterations of con-sciousness, with or without signs of localization. Numerous classifications have been made mainly concerning the seriously cranially traumatised patient and particularly those in coma (Vigouroux and Guillermain 1986). However, whether these classifications refer to a progressive deterioration in the state of vigilance, or to clinical sign scales as in the Glasgow scale, or to signs of axial involvement from rostrocaudal destruction, they take no account of the lesions. Thus, even if a contusion syndrome has been described (Basauri *et al.* 1973; McLaurin and Helmer 1965), the clinical

point of view does not allow cerebral contusions–lacerations to be distinguished from other brain trauma complications. Besides, owing to modern
methods and means of transport and treatment, most of the seriously
cranially traumatised patients are now hospitalised, intubated, ventilated,
with a neuro-vegetative disconnection. This treatment, which may be continued for a certain length of time, makes all interpretation of the clinical
examination hazardous. It is therefore only from anamnesis and during
suspensions of therapeutic treatment that the clinical condition can be
determined.

Disturbances of Consciousness

These are constantly recorded, and are found to be present in all the
series, with a frequency around 90%. They appear either at the time of the
trauma, or after a free interval of variable duration in the form of coma or
drowsiness. For example, in our series (Vigouroux and Guillermain 1981)
we observed primary disturbances in 74% of cases (50% comatose, 24%
drowsy) and delayed altered consciousness in 22% of cases. This notion of
a free interval is underlined by most authors, with a frequency ranging from
15% (Cohadon et al. 1973) to 44% (Casella et al. 1967). Our series recorded
22% (184 out of 833 cases). Among these patients only 21% had an
associated extracerebral hematoma. Consequently, noting such a possibility must not a-priori exclude a possible cerebral lesion. In fact, in nearly half
the cases, the interval preceding aggravation is not clear cut; that is why we
prefer to say "free" rather than "lucid" interval.

In nearly 10% of cases, the state of consciousness can remain normal
throughout the evolution. Since the systematic use of the CT scan, these
states evolving without alterations of consciousness, have been more frequently recorded whatever the gravity of the lesions (Pierron et al. 1972).
For example, in our study of two comparative series, we noted these states
evolving without disturbances of consciousness in 4% of cases prior to the
use of the CT scan, and in 15% of cases afterwards (Vigouroux and
Guillermain 1983).

Signs of Localization

These can be found in more than half the cases (Basauri and Rocamora
1968; Da Pian et al. 1967; Houdart et al. 1968; Rusu 1972; Vigouroux et al.
1982). The examination most often reveals a pyramidal syndrome evidenced by a motor deficit, or a unilateral mydriasis, or both associated,
resulting in an "altern pseudo-syndrome". Their proportion varies according to different authors. In our series (Vigouroux and Guillermain 1981),

out of 833 patients, 557 (67%) presented with signs of localization: 44% pyramidal syndrome, 28% mydriasis, 28% both. Bravais–Jacksonian seizures are considered to be a good sign of focal lesions (Houdart *et al.* 1986), but are only found in 5% (Vigouroux and Guillermain 1981) to 8% (Casella *et al.* 1967) of cases. These signs are not specific, but two facts are worth noting:

– The signs are most often delayed, though they are early as regards trauma. For example, our series (Vigouroux and Guillermain 1981) recorded 41% of signs existing as from the time of trauma, and 59% of signs after a delay. Among the latter signs, 53% appeared within 48 hours, and 70% within three days. This may be because it is sometimes difficult to correctly detect hemispheric symptomatology in a comatose patient, all the more so when the patient has neuro-vegetative disconnection.
– The localizing value of the signs seems less dependable than in hematoma collections (Basauri and Rocamora 1968) as with 20% (Vigouroux and Guillermain 1981) to 30% (Basauri and Rocamora 1968) of the injured, signs do not correspond with the site of the lesion. Finding a contralateral mydriasis or an ipsi-lateral motor deficit also constitutes a sure sign of gravity.

More specific signs of the topography of the lesion can be observed when the acute phase has improved or when the state of consciousness has remained normal: disturbances of language, mental changes, various sensorial disturbances particularly in the visual area. When these deficits persist for a long time and become sequelae, they constitute a definite handicap for the patient.

Diffuse Neurological Signs

Their prognostic value varies. Some signs have hardly any particular incidence – for example, the meningeal syndrome which is frequent in this type of lesion: 40% (Vigouroux and Guillermain 1981) to 85% (Casella *et al.* 1967) of cases, some of which (Schneider 1966) might be characteristic of a localised cerebral hemorrhagic lesion. Other signs reveal the gravity of the trauma: signs of axial involvement are present with half the severely head injured. Some of the signs, like abnormal reflexes ascribable to the brain stem or problems with muscle tone, are criteria used in appreciating the gravity of traumatic coma, and their prognostic value is known. We will not dwell on neuro-vegetative signs, as careful observation of the vital features has for a long time been considered essential in the surveillance of the cranially traumatised.

Clinical Synthesis

Among the clinical entities, some have simply a speculative interest, as they cannot be ascribed real diagnostic values: anatomo–clinical forms. Others are considered as being more interesting, as the indications for surgery largely depend on them: evolutive forms:

1) Anatomo–clinical Forms

They are now of limited interest, since the CT scan permits a precise morphological diagnosis.

– We will not insist on topographic forms, as in the acute phase, symptomatology related to a frontal, temporal, parietal or occipital lesion is in most cases hidden by alterations of consciousness. It is only when the state of consciousness is normal that it may be possible to define this symptomatology.
– Anatomical forms: according to the importance of the tissue destruction, of the hemorrhage, or of the oedema, three anatomo–clinical forms can be differentiated through the possible confrontation between the clinical and surgical data (Vigouroux and Guillermain 1981; Vigouroux *et al.* 1982), the latter being corroborated from information provided by the CT scan:

Cerebral contusion: it is not possible to draw up a single clinical pattern of cerebral contusion, as all degrees of severity are possible. However, one form, contusion oedema, should be specially considered: on examination the brain appears red, swollen, turgid, with superficial ecchymoses and hemorrhages originating from capillaries. In most cases the patients are in a serious state of primary coma showing focal signs, if they exist, at the first examination. Signs of axial suffering are frequent. This form appears to be particularly serious: thus in our series (Vigouroux and Guillermain 1981) the death rate was about 82%.

Cerebral laceration: on examination, the brain has the appearance of necrotic infarcted pulp with blood clots. In most cases the trauma caused initial coma which lasted a variable period of time. Improvement in the state of consciousness can be observed, however with drowsiness and agitation persisting. Clinical examination reveals discreet but often discordant focal signs, disturbance of language often being mistaken for confusion.

Bravais–Jacksonian crises are relatively frequent. The evolution is marked in the first 2 or 3 days by further depression of consciousness and with focal signs becoming more distinct. It seems that we may envisage a particular form of polar and basal (frontal and temporal) laceration, having a much longer evolution. Indeed, this evolution may extend over several days, even weeks, with the development of an intracranial hypertension syndrome and papilloedema in patients where dowsiness persists without focal symptomatology. The gravity of these forms seems to depend on the importance of tissue damage since we noted (Vigouroux and Guillermain 1981) a death rate of 8% with superficial lesions, 43% with lacerations, and 62% when there was actual rupture of the cortex.

Hematoma contusions: here, the hemorrhagic factor prevails rather than the destructive one. On examination, there is a laceration zone, often restricted and adjacent to a hematoma cavity, the hemorrhage having extravasated into the softened parenchyma. The symptomatology is identical to that of contusions or lacerations, but in some cases, trauma is followed by a time gap of variable duration, often prolonged, with focal symptomatology associated with signs of intracranial hypertension, a pattern which is reminiscent to that of spontaneous intracerebral hematomas. This may suggest pre-existing vascular fragility or a vascular malformation which was not appreciated. Sometimes, particularly with old people, benign trauma raises the question whether the loss of consciousness may have been the causal factor. These forms appear to have a better prognosis than the previous ones: indeed we noted a 35% death-rate (Vigouroux and Guillermain 1981).

It seems hazardous to try and establish anatomo–clinical correlations, as all the clinical patterns may be observed whatever the type of lesion, and this especially relates to multiple lesions, or if there are associated extra-cerebral hematomas.

2) Evolutive Forms

The evolutive modalities of disturbances of consciousness are considered to be pre-eminent by all authors, as the prognosis and the surgical indications, will partly depend on this evolution. While some authors (Casella *et al.* 1967; Cohadon *et al.* 1973; Houdart *et al.* 1968) consider two groups of patients: those whose altered consciousness occurs at the onset and persists, and those patients who deteriorate whatever the initial state of consciousness; we think that there are four evolutive patterns (Schisano *et al.* 1975) that should be separately recognised during the early days following the trauma:

– *extremely serious forms*: these concern injured patients with an almost moribund initial state on hospitalization leading to rapid death;
– *serious forms*: these concern patients who are in coma from the trauma, and this state tends to vary little, either remaining unchanged or aggravating progressively deteriorating;

– *less serious forms*: here, alterations of consciousness evolve, with or without a free interval and result in a state of coma. Whatever the degree of damage they induce in the injured, deterioration is progressive in two out of three patients, but may also be extremely sudden in one out of three patients. In our study, the type of deterioration appeared to be related to the site of the lesion: temporal lobe lesions tend to worsen progressively whereas sudden deterioration is more frequently noted with frontal lobe lesions (Vigouroux and Guillermain 1981). Most injured patients become worse during the first three days, most commonly in the first 24 hours;
– *benign forms*: disturbances of consciousness are not alarming. They pertain to patients whose primary alterations of consciousness improve rapidly, or patients who, from the onset, present a state of drowsiness which persists without substantial variation, or in patients presenting with no disturbance of consciousness at any time.

Each of these forms poses different therapeutic problems. Indeed, alteration of vigilance influences the prognosis and death-rates increase as the clinical state deteriorates. For example, taking into account the state of consciousness on intervention, we noted death-rates of 85% with patients in coma III–IV, of 50% with patients in coma I–II, of 6% with patients who were in a state of drowsiness or were lucid (Vigouroux and Guillermain 1981). When the lesion is minimal or moderately important, whatever its evolutive form, treatment may remain medical while maintaining clinical and CT scan surveillance. With more serious lesions, except for extremely severe forms, surgery appears to be indicated in most cases if the patient is in a grave state or if his condition deteriorates. Thus the only problem is to choose the best moment for surgery: in emergency, very early or after a delay.

With benign forms, what attitude should be adopted with patients whose clinical state is satisfactory, but who are likely to deteriorate abruptly at any moments? Should surveillance be maintained with the risk of intervening too late, or on the contrary, should patients be operated on, and if so, when?

References

1. Basauri L, Rocamora RJ (1968) Traumatismo encefalo craneano. Revision de 500 casos. Neurocirugia 26: 154
2. Basauri L, Palma A, Henriquez C (1973) Traumatic temporal lobe syndrome. 5th Int Congr of Neurological Surgery, Tokyo 1973, vol 7, p 13
3. Botterel EH, Stewart W (1947) Disruption of temporal or frontal lobe as a cause of secondary coma following head injury. Harvey Cushing Society

4. Bues E, Schmidt H (1961) Neurologisches Bild der frischen gedeckten traumatischen Hirnschäden. Bruns-Beitr Klin Chir 203–265

5. Casella E, Chiapetta F, Chiasserini A, Gazzeri G (1967) Considerazioni clinico-statistiche su 150 casi di lacerazione cerebrale. Minerva neurochir 11: 154

6. Cohadon F, Richer E, Castel JP *et al* (1973) Aspects cliniques et angiographiques des lésions parenchymateuses fronto-temporales d'origine traumatique. Neuro-Chirurgie 19: 417

7. Columella S (1973) Traumatic brain laceration has a new and independent nosological entity of neuro-surgical pathology. Neurocirugia 31: 9

8. Da Pian R, Dalle Ore B, Bricolo A *et al* (1967) Lacerazioni cerebrali traumatiche. Considerazioni su 190 casi operati. Minerva neurochir 11: 147

9. Fasano VA (1973b) Aspects anatomo-cliniques des hématomes sous-duraux et des attritions cérébrales post-traumatiques. Neuro-Chirurgie 19: 430

10. Feld M, Daum S, Minuit P (1955) La contusion-hémorrhagie du lobe temporal dans les traumatismes crânio-cérébraux fermés. Indications opératoires. Neuro-Chirurgie 1: 333

11. Geuna E, Pagni CA, Caneschi S (1967b) Studio di 135 casi di focolai lacero contusivi edematomi intra-cerebrali traumatici trattati chirurgicamente negli anni 1960–1966. Minerva neurochir 11: 127

12. Gruner V, Krenn J, Kutscha *et al* (1965) Quelques sonnées sur le pronostic des contusions cérébrales. Neuro-Chirurgie 15, 4: 275

13. Guillermain P (1970) Les lésions cortico-sous-corticales localisées au cours des traumatismes crâniens fermés récents (à propos de 234 observations). Thèse Médecine, Marseille

14. Gurdjian ES, Webster JE, Lissner H (1958) Mechanism of scalp and skull injuries; concussion, contusion and laceration. J Neurosurg 15: 125

15. Houdart R, Cophignon J, Hurth M *et al* (1968) Les traumatismes crânio-cérébraux. E.M.C. Neurologie 17585 A 10

16. Lazorthes G (1973) Hématome sous-dural aigu et attrition cérébrale post-traumatiques. Table Ronde, Neuro-Chirurgie, 19, 5: 415–490

17. Liguori R, Troisi F (1966) Focolai cerebrali traumatici lacero-contusivo-emorragici chiusi. Rass int Clin Terap 47: 3

18. MacLaurin R, Helmer F (1965) The syndrome of temporal lobe contusion. J Neurosurg 23: 296–303

19. Obrador S, Gomez J, Garcia M (1968) Contuson destructiva y edematosa del lobulo temporal en los traumas craneales cerrados. Revta esp.O.N.O. Neurocirugia 25: 39

20. Pierron D, Georges B, Ouahes O *et al* (1981) Tomodensitométrie et hématomes intra-crâniens posttraumatiques sans manifestation clinique. Neurochirurgie 27: 213–216

21. Rusu M Localized traumatic meningo-cerebral injury (1972) Neurochirurgia 6: 209

22. Schisano G, Schonnauer M, Cimino R *et al* (1975) Space occupying contusions of cerebral lobes after closed brain trauma. 6th Int Congr of Neurological Surgery, Sao-Paulo, vol 6, p 77

23. Schneider RC (1966) Serious and fatal neurosurgical football. Clin Neurosurg 12: 226
24. Stender A, Schulze A (1966) The surgical treatment of space-occupying contusions and intra-cerebral hematomas after blunt cerebral trauma. Excerpta Med Int Congr Ser, N° 110, p 231
25. Teasdale G, Galbraith S (1981) Acute traumatic intracranial hematomas. Prog Neurol Surg, vol 10. Karger, Basel, pp 252–290
26. Tönnis W, Friedmann G, Schmidt-Wittkamp E *et al* (1965) Les hématomes intra-crâniens traumatiques. Geigy, Basel
27. Vigouroux RP, Guillermain P (1981) Posttraumatic hemispheric contusion and laceration. Prog Neurol Surg, vol 10. Karger, Basel, pp 49–163
28. Vigouroux RP, Baurand C, Guillermain P *et al* (1982) Traumatismes crânio-encéphaliques. E.M.C. Paris, Neurologie 17585 A 10, A 15, A 20, 10
29. Vigouroux RP, Guillermain P (1983) Surgical indication in post-trauma brain contusion and laceration. Advances in Neurotrauma. Excerpta Medica Int Congr Ser 612: 128–134
30. Vigouroux RP, Guillermain P (1986) Classifications en traumatologie crânio-cérébrale grave. Le coma traumatique. Liviana Press, Padova, pp 7–64

Traumatic Brain Swelling and Brain Edema

K.E. Richard

Neurochirurgische Klinik der Universität zu Köln
(Federal Republic of Germany)

With 8 Figures

Contents

Introduction

In a similar manner to other tissues of the human body, the brain responds to an injury with disturbances of cerebral metabolism and blood flow, leading to hyperemia and edema. Increases in cerebral blood volume and water content of brain tissue, or a combination of the two, lead to brain bulk enlargement. The cause of this phenomenon is still an object of intensive research activity, however, even the terms and definitions for this phenomenon have not been already established to date.

Reichardt (1905), referred to the "dry" cut surface as "brain swelling", since fluid could not be demonstrated histologically (Zülch 1967). If, on the contrary, the cut surface was "wet" and the presence of fluid in the free spaces between the tissue elements was confirmed histologically, he termed this condition "edema".

"There has been much discussion as to whether this distinction between brain edema and brain swelling is of major importance for neuropathology and neurosurgery" (Zülch 1986).

Using the present-day diagnostic aids, such as computed tomography (CT) and nuclear magnetic resonance tomography (NMR), a more concise definition of the term "brain swelling" is possible. It is characterized by a local or diffuse increase of brain tissue density, which can be recognized by an absence or compression of the intracranial fluid spaces, such as ventricles, cisterns and the subarachnoid spaces and by a slightly elevated density of the brain tissue (Zimmerman *et al.* 1978).

The term "brain edema" is related to an increase of brain tissue volume based on perifocal or diffuse increase of brain water content with a decrease of tissue density on the CT (Langfitt 1983; Clasen 1986).

Diagnosis and Pathogenesis of Traumatic Brain Swelling and Brain Edema

1. Traumatic Brain Swelling

1.1. Diagnosis

Computertomographic diagnosis of *brain swelling* (BS) is based on the following features:

– Brain tissue lacks normal structure and has a homogeneous appearance with isodense or slightly elevated hyperdense values,

– Lateral ventricles and third ventricle, perimesencephalic cisterns and subarachnoid spaces are compressed or are absent,
– Tissue absorption is slightly elevated in comparison to normal brain (Lanksch 1982).

Traumatic brain swelling is associated with an increase in the *CT density of white matter* of about 3–4 Hounsfield units (HU) (Bruce 1981; Ito *et al.* 1985). Clasen (1986) however, believes that vascular congestion could not explain the increase of white matter density, because a doubling of the white matter substance should increase its density values only from 30.0 to 30.7 HU. Ito *et al.* (1985) differentiated 3 *grades of acute brain swelling*: (1) mild: ventricles and cisterns were narrower on the CT scan in the acute stage in comparison to those scans taken at 3 weeks or more after head injury,

(2) moderate: as above and in addition, the sylvian fissure and cortical gyri were not visualized,

(3) marked: as in (1) and (2) and in addition, the perimesencephalic cisterns were not visualized.

1.2. Pathogenesis

The cause of these CT features of brain swelling is still a matter of discussion. Lanksch *et al.* (1982) assume, that the state of brain swelling is a precursor of early brain edema, which is not yet detectable. This causes a rapid increase in lipid content, which could be responsible for maintaining normal or slightly elevated density values. This assumption, however, has not yet been confirmed by means of biochemical analyses of tissue samples.

Other authors suggest, that increased cerebral blood flow may play an important role. Several series of measurements of regional cerebral blood flow (CBF) in children and adolescents with craniocerebral injury as reported by Zimmerman (1978) and Bruce (1981), showed a distinct increase of CBF up to 75 ml/100 g/min. Repeated CBF studies after resolution of diffuse swelling in those patients who survived, revealed a decrease in CBF in all patients. Thus, a true hyperemia was present in association with the pattern of diffuse swelling (Bruce 1981). Kuhl *et al.* (1980) used positron emission computed tomography (ECT) following injection of technetium-99-m labelled red cells in *cerebral blood volume studies* in 30 patients after head injury. Their studies revealed a mean value for cerebral blood volume (CBV) of 4.34 ml/100 g for the total posttraumatic period. However, in 5 adults with predominant unilateral lesions, an average reduction of 15–36% in mean CBV and CBF respectively, was observed in the early posttraumatic phase. In 3 children with diffuse brain swelling, there was an average early mean CBV reduction of 12%. The authors admit, however, that their method has limitations, because the data represent not only blood volume in capillaries, but are also strongly influenced by large-vessel volumes, mainly in the veins of the surface of the brain. Furthermore, the spatial resolution of this ECT method is relatively poor, and the local values for cerebral hematocrit have not been determined accurately in the small structures of the normal brains.

In 1984 Obrist *et al* described a xenon-133 intravenous injection method in 75 adult patients with closed head injuries and found that 55% of the patients developed an *acute cerebral hyperemia*, which they defined as excessive blood flow relative to the cerebral metabolic rate for oxygen ($CMRO_2$). The hyperemia lasted an average of 3 days. 45% of the patients had consistently subnormal CBF values together with a low CO_2.

In 45% of the patients with acute hyperemia, the CT scan revealed signs of diffuse cerebral swelling, such as compression or obliteration of the ventricles and basal cisterns, but these findings were observed in only 19% of the injured patients with reduced CBF (Langfitt 1983).

Yoshino *et al.* (1985) performed "dynamic" computerized tomography in 42 patients with acute head injury, in order to evaluate the hemodynamics and the nature of *diffuse brain bulk enlargement* (BBE). They concluded that BBE following head injury originates from both an increase in CBV, as well as acute brain edema. In those patients who survived, the hyperemic state predominated; whereas in those patients who died, rapidly occurring increase in brain water content was thought to be responsible.

Bullock *et al.* (1986) correlated CT scan density with *specific gravity* of the *white matter* in 60 patients with severe head injury and found a good correlation between these values. 17 patients had CT density values of 36 HU or greater together with high specific gravity values for white matter. These findings were interpreted as probably representing a condition of "diffuse brain engorgement". 5 patients with diffuse injuries had CT density values of 30 HU or less, in association with a white matter specific gravity considerably lower than the normal value. These patients were thought to represent "diffuse brain edema".

To explain the pathogenesis of brain swelling and the associated ICP increase, Langfitt and coworkers developed the *concept of cerebral vascular congestion* caused by cerebral vasomotor paresis. According to this hypothesis, following a severe experimental head injury, a progressive rise in ICP ultimately causes vasomotor paralysis. They proposed that the resulting vascular engorgement rather than edema, was the primary cause of both brain swelling and the increase in ICP during the early posttraumatic period.

According to Langfitt and Obrist, important *factors connected with hyperemia* are: (1) a decrease in cerebral vasomotor tone, which is suggested by the consistent findings of *impaired CBF autoregulation* in response to changes in blood pressure. Therefore hyperemia may be triggered by a large swing in systemic arterial pressure which breaks through the upper limit of CBF autoregulation. Profound increases in systemic arterial pressure, often accompanied by ICP elevations, have been observed immediately following experimental trauma (Langfitt 1966). This notwithstanding, a total loss of vasomotor reactivity seems unlikely, since head injured patients retain some degree of responsiveness to changes in $PaCO_2$. Defective autoregulation and preserved CO_2 reactivity have been described and termed "dissociated vasoparalysis" (Paulsson 1972). (2) the most likely cause of posttraumatic hyperemia is *metabolic acidosis of the brain* brought about by an increased lactic acid production (Enevoldsen *et al.* 1976).

2. Traumatic Brain Edema

2.1. Diagnosis

Brain edema (BE) may be defined as an increase in brain volume due to an *increase in brain water content* (Fishman 1955).

Computerized axial tomography (CT) with high resolution, permits earlier detection and delineation of BE, and enables one to differentiate between brain swelling and brain edema (Langfitt 1983; Ito *et al.* 1986).

Using *computerized tomography*, one can differentiate two forms of edema: (a) *focal edema*, which develops outside of the area of contusion. This form is more of the vasogenic type. (2) *diffuse* form of traumatic *edema* as a consequence of an ischemic lesion.

2.2. Pathogenesis

Like most body tissues, the brain responds to trauma by acute swelling of brain tissue. The assumption that an increase in cerebral blood volume (CBV) might be a primary cause of the acute brain swelling (Langfitt *et al.* 1966; Lewelt *et al.* 1980) has received more attention in this regard than the possible role of brain edema.

Miller and Corales (1982) found no increase in *grey and white matter water content* during elevated ICP at 5 minutes and 30 minutes after trauma. Their experimental model was based on concussive brain injury of high intensity. Using the same experimental setting, Becker (1986) however, measured significant brain edema at a later stage, namely 24 hours after the impact. With central injury, the edema was maximum in the brain stem.

The increase of water content was confined to the white matter. With lateral injury, the greatest region of edema was measured in the impacted hemisphere.

Tornheim and McLaurin (1981) were able to measure a significant *decrease of specific gravity*, corresponding to an increase in the tissue volume in the contused hemisphere at 30 minutes after head injury. They used another mechanical impact model in which the unilateral contusional injury was provoked in cat brains. Data at 1 and 2 hours after trauma were similar, with a trend towards an increase in tissue volume for subcortical and deep white matter. At 6 hours following cranial impact, density of the deep white matter showed a substantial increase in tissue volume and in the caudate nucleus significant edema was demonstrated for the first time. The authors conclude that BE contributes to early brain swelling following closed head injuries.

Recently, Sakamoto *et al.* (1986) investigated the *contribution of BE to acute brain swelling* using a conventional balloon compression model. Some 50 minutes after deflation of the balloon, the water content of white matter was increased in both hemispheres and was associated with an increase in ICP. These changes occurred simultaneously with a deterioration in autonomic nervous and EEG-activity.

Traumatic cerebral edema is not a single entity. It includes both vasogenic and cytotoxic components.

Klatzo has classified BE into 2 types, depending on the underlying basic mechanisms (Klatzo 1967):

(1) *vasogenic*, in which the main event is a regional increase in cerebrovascular permeability to plasma contents, including water, which leads to an extracellular spreading of edema fluid.

(2) *cytotoxic*, which is characterized by an abnormal uptake of water by various cellular elements of the brain parenchyma.

Naruse *et al* (1982) studied the *water content* of normal and *edematous brain tissue* in rats using the pulse nuclear magnetic response (NMR) technique. They found significant differences between the 2 types of BE: quantitative analyses of longitudinal (T1) and transverse (T2) relaxation time values as regards water content over time, demonstrated that prolongation of T1 was related to the volume of increased water in tissues of both edema types. However, the two phases of T2, which reflect the distribution and the content of edema fluid in grey and white matter, revealed a significant difference in both types of edema.

2.2.1. Vasogenic Brain Edema

Vasogenic brain edema develops in the periphery of an area of direct damage to brain tissue. This causes disruption of neurones, glial cells and their processes, as well as of blood vessels. This in turn leads to a zone of ischemic necrosis. Such areas are bordered in turn by an area of dysfunctional cerebral vessels.

These demonstrate both abnormal permeability at the capillary level and loss of normal physiologic regulation at the arteriolar level. This type of brain damage is simulated in the experimental models of cryogenic or of fluid percussion injury.

The most frequently used *model of edema* of the traumatized brain is local cold injury to the exposed cortex. This *cryogenic model* (Klatzo et al. 1967) has the advantage of high reproducibility, but has the disadvantage of not resembling the clinical situation in traumatic BE. On this account, many investigators (Becker, Marmarou, Miller et al.) would favor the *fluid percussion method* (Ommaya et al. 1964).

Using this latter model, Povlishok and Kontos (1982) evaluated traumatic intraparenchymal vascular changes. Using SEM-and TEM-electron microscopic analysis, they recorded widespread endothelial changes throughout the intraparenchymal vasculature, associated with temporary alterations in the blood brain barrier (BBB). These were characterized by endothelial balloons and craters, constituting foci of endothelial damage with mechanical disruption of the endothelial tight junctions. This pathogenic alteration of BBB results in an increased vascular permeability, which allows escape of plasma or plasma filtrate into the surrounding extracellular space.

The possible causes of transendothelial passage of serum constituents in BBB-impairment which in turn lead to formation of vasogenic BE, have been discussed by Houthoff (1987) and include the following: (1) opening of interendothelial junctional areas, (2) vesicular transport, (3) transendothelial channels, (4) regressive endothelial changes with leaky membranes.

In this concept, the most important fact is, that the extravasation of the edema fluid is confined to the area of the primary focus of injury (Reulen and Tsuyumi 1982).

An excellent survey on the formation, spread and natural resolution of vasogenic brain edema, based on the cryogenic model, has been made by Reulen (1982).

2.2.2. Cytotoxic Brain Edema

The second form of water accumulation in brain tissue is intracellular, and this can be considered to be a passive process accompanying the cytoplasmic dissolution of dying cells. A general understanding of the *nature of cytotoxic BE* has not yet been reached (Kempsky 1986).

The basic *cause of cytotoxic BE* is an energy deficit, resulting in improper functioning of the cellular Na/K – pump with inadequate pumping of Na and water out of the cells. For the posttraumatic situation, this is assumed to be the case both in hypoxia, and in the initial stage of ischemia (Go 1982). The cells which are affected are glial elements, especially astrocytes.

One to 3 minutes after the onset of complete *cerebral ischemia*, the control of cell volume is completely lost. Electron microscopic investigations and impedance resonance studies of extracellular tracer substances (Kempski 1986) revealed, that the extracellular space shows a volume reduction of at least 50% within a few minutes. The normal extracellular potassium and sodium concentrations are reversed and for one or two hours a marked increase in tissue osmolality in the grey matter of the affected territory is observed which in turn may provide a basis for ischemic brain edema (Hossmann and Matsucka 1983).

3. Mediators of Brain Edema

In recent years, extensive research has provided evidence, that the early onset of ischemic cell swelling *in vivo* is likely to result from an accumulation of toxic compounds in the extracellular space (Baethmann 1978; Kempski 1986; Unterberg *et al.* 1986).

In Table 1 the *chemical mediator compounds* present in plasma or which are activated from damaged brain tissue, and which in turn may be involved in the development of secondary brain damage such as brain edema, are summarized.

These mediator compounds are believed to be formed, or to become released primarily in areas of brain tissue, *e.g.* the focal necrosis of cerebral

Table 1

Mediator compounds of secondary brain damage
(acc. to Baethmann 1986)

– proteolytic, lysosomal enzymes
– active peptides: kallikrein-kinine-system
– biogenic amines: serotonin, histamine
– neurotransmitters, aminoacids: glutamate
– fatty acids: arachidonic acids and metabolites, *e.g.* prostaglandines, leucotrienes
– oxygen free radicals
  eicosanoids

---

contusion, or to enter the brain parenchyma via the intravascular compartment.

Such mediators might

- enhance damage to the BBB, thereby promoting influx of vasogenic edema,
- induce circulatory disturbances, such as impairment or loss of the cerebral autoregulation and vasocongestion resulting in cerebral ischemia,
- induce secondary cell swelling, *i.e.* cytotoxic edema.

Swelling of glial cells is regularly found in areas with primary vasogenic edema, probably resulting from release of mediator substances. The kallikrein–kinine system, glutamate and arachidonic acid appear particularly important in this context, as the development of brain edema has been demonstrated after cerebral administration of these compounds (Unterberg 1987).

Recently Baethmann *et al.* (1983) demonstrated experimentally, that compounds with neurotoxic properties, such as the *kallikrein–kinine system* and glutamic acid, are activated or released in focal and perifocal edematous brain tissue. Kinine-concentrations of only 10^{-6} M were found to induce porosity of the BBB. Pores of at least 11 Å in diameter were formed, which are large enough to allow penetration of electrolytes and water through damaged areas of the BBB. The increase of glutamate in the extracellular space raises Na-permeability and causes an influx of water into the intracellular compartment, with resulting cellular edema if the compensatory Na efflux mechanisms fail.

Both the release of glutamate and the formation of kinins could precipitate an increase in ICP due to the formation of cytotoxic cell swelling and vasogenic edema.

Frequency and Temporal Development of Traumatic Brain Swelling and Brain Edema

1. Frequency

As regards the frequency of posttraumatic brain swelling and brain edema in patients with severe head injuries treated in neurosurgical units, the incidence quoted in the literature ranges from 9–75% for posttraumatic brain swelling and from 12–27% for posttraumatic brain edema.

This discrepancy is due to the fact that *different definitions* and selection criteria are used to describe these entities. For instance, the lower incidence of brain swelling reported by some authors, refers only to pure forms of diffuse brain swelling without additional lesions, such as hematoma or contusional hemorrhage.

Lobato et al. (1983) outlined basic patterns of traumatic and posttraumatic lesions which have physiopathological and prognostic significance. Their findings were based on the pathological changes noted in serial CT scanning of 277 patients with head injuries. However, their classifications failed to distinguish between brain swelling and brain edema. BE is designated only as a "hypodense" swelling.

In 273 consecutive cases of severe head injury treated in 1985/86, we correlated *typical findings* in the CT scans with frequency of occurrence and patient outcome (Richard et al. 1989). Based on the *CT-scan*, we would like to propose the following *classification* (Fig. 1):

0 : No visible alteration
1 : Brain swelling
   (According to Zimmerman et al. (1978) BS was assumed when the perimesencephalic cistern, the ventricles or the subarachnoid spaces of the brain surface were compressed or effaced)
2 : Contusional hemorrhage (CH) or hematoma without signs of BS
3 : CH or hematoma with signs of BS
4 : CH or hematoma with perifocal BE or local postischemic BE
5 : CH or hematoma with BS and BE

In the majority of the patients the CT-scan was performed within the first 3 hours after trauma. The remainder of the patients firstly were scanned between 4 to 24 hours after trauma. Table 2 shows the percentages of patients regarding the severest CT findings.

15% of the head-injured patients had *no pathological findings* on CT, 16% developed *diffuse BS without additional lesions*, but only 4% developed *contusional hemorrhages or hematomas without any other alteration*. In 32% of the patients *CH* or *hematoma* appeared together *with* signs of *BS*. Similarly a further 31% had *CH or hematoma* together *with* signs of *BE and BS*. Only 2% of the patients suffered from CH or hematoma with BE but without signs of BS.

Thus, the *most frequent findings* were contusion or hematoma with accompanying BS and/or BE.

The frequency of *isolated BS* was highest in children and juveniles (26%), decreasing to 17% and 14% in young and in middle-aged adults respectively, and being only 5% in the elderly. However, BS associated with intracranial bleeding was more frequent in the elderly than in the younger patients. CH or hematomas associated with signs of BS and BE, were more frequent in the elderly than in children or in young adults.

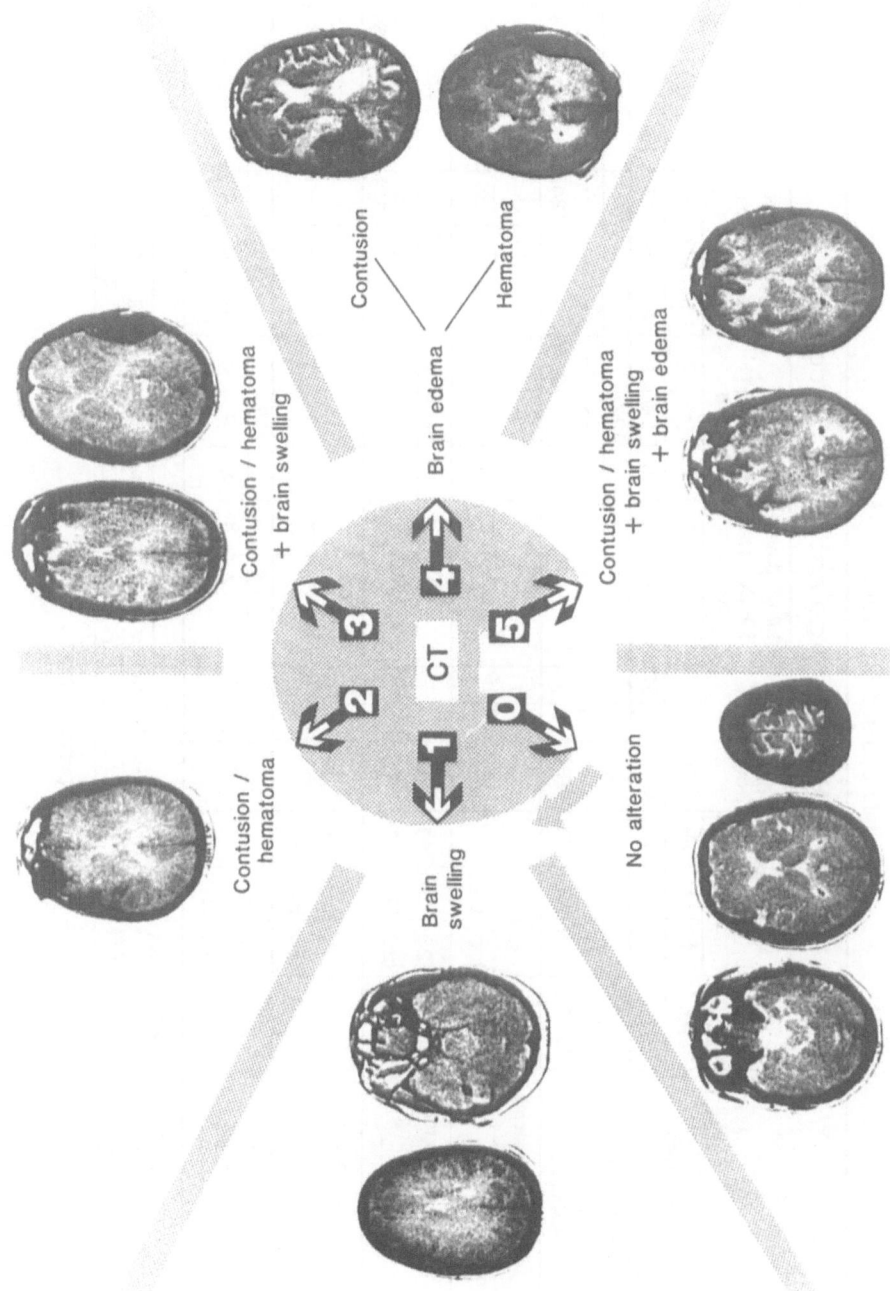

Fig. 1. Findings on CT-scan after severe head injury (according to Richard *et al.* 1989)

Table 2. *CT-findings in Patients After Severe Head Injury.* Frequency, age and early outcome of patients in the differentiated CT-groups (0–5)

GROUPS	CT-Findings	No. of Pts	All Age Groups N:273 ↑	(↑)	✝	Within 2 days ✝	Age x̄	SD	Child + Juven. [0–20 yrs] N:54 ↑	(↑)	✝	Adults [21–40 yrs] N:98 ↑	(↑)	✝	Middle Aged [41–60 yrs] N:71 ↑	(↑)	✝	Seniors [>60 yrs] N:50 ↑	(↑)	✝
	(totals)	273	93	107	73	32 [44]			19	25	10	37	34	27	27	29	15	10	19	21
5	Contusion/Hematoma + Brain Swelling + Brain Edema	85 [31%]	14 [18]	47 [55]	24 [28]	3 [13]	43	21	1	10 [26]	3 [21]	6	13 [27]	7 [27]	7	15 [41]	7 [24]	–	9 [31]	7 [44]
4	Contusion/Hematoma + Brain Edema	5 [2%]	3 [60]	2 [40]	–	–	54	19	–	–	–	1	–	–	2	1 [28]	–	–	1 [33]	–
3	Contusion/Hematoma + Brain Swelling	88 [32%]	23 [26]	24 [27]	41 [47]	28 [68]	39	21	5	5 [30]	6 [38]	7	9 [36]	19 [54]	8	6 [28]	6 [30]	3	4 [33]	10 [59]
2	Contusion/Hematoma	10 [4%]	3 [30]	5 [50]	2 [20]	1	59	24	1	–	–	1	1 [19]	–	–	2 [13]	–	1	2 [6]	2
1	Brain Swelling	43 [16%]	19 [44]	20 [47]	4 [9]	–	29	18	5	7 [24]	1 [8]	8	9 [19]	1 [6]	4	4 [13]	1 [44]	2 [6]	–	1 [33]
0		42 [15%]	31 [74]	9 [21]	2 [5]	–	37	22	7 [19]	3	–	14 [16]	2	–	6 [10]	1	1	4 [16]	3	1

Patient's Age and Early Outcome

↑: fully recovered; (↑): disabled; +: dead; ( ): percentages

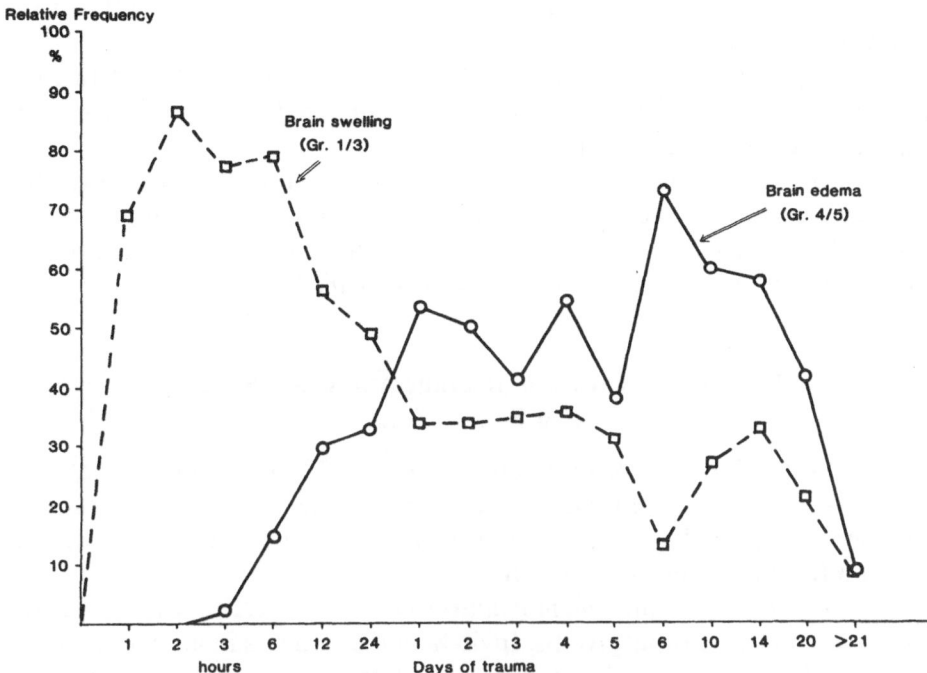

Fig. 2. Relative frequency of brain swelling and brain edema on CT-scan in the early posttraumatic course

2. Temporal Development

Signs of *brain swelling* appeared in a relatively high percentage of patients (70%) within one hour of trauma. CT-scan performed at 2 hours after injury revealed brain swelling in 87% of the patients (Fig. 2).

The relatively high percentage of BS without BE decreased after the 6th hour.

Ito *et al.* (1985) analyzed the temporal course of the severity of acute brain swelling after head injury, as seen on CT imaging. Brain swelling appeared immediately after head injury and became less severe 1–2 days after trauma.

Obrist *et al.* (1984) found that hyperemia is of a transient nature with an average duration of 3 days and a range of 1 to 8 days, but they differentiated between 2 distinct CBF patterns involving an initial and a delayed hyperemia.

Signs of *brain edema* in our patients were first evident between the 3rd and the 6th hour after trauma (Fig. 2). The relative percentage of BE increased sharply after this time, reaching 33% after 24 hours, 54% after 48 hours and 73% up to the 6th day. After this peak, the relative percentage of

BE decreased to 59% at the end of the 2nd week, and reached 42% at the end of the 3rd week.

As mentioned above, Tornheim and McLaurin (1981) found in their experiments, that significant *brain edema* appeared at the earliest some 6 hours following the cranial impact.

Proton nuclear magnetic resonance – NMR – studies have shown that the *maximum* degree of edema following the vasogenic cold injury is present after 24 hours, whereas cytotoxic edema following application of the trietyl-cin reaches its maximum on day 7 (Naruse *et al.* 1982).

Intracranial Pressure in Traumatic Brain Swelling and Brain Edema

In 63 out of the above-mentioned group of 273 head injured patients, *ICP* was continuously measured over an average period of 7 days. Our comments on the CT findings refer exclusively to patients in group 1 (n:10), group 3 (n:19) and group 5 (n:34).

Table 3 shows the *maximum epidural pressures* (EDP) measured at the time of CT in the various groups. In 40% of the patients in group 1 (*general brain swelling*) ICP rose above 20 mmHg, but in no case above 40 mmHg.

As for the patients in group 3 (*CH/hematoma with BS*), the EDP was raised above 20 mmHg in 63%, above 40 mmHg in 32% and above 60 mmHg in 11%.

In group 5 (*CH/hematoma with BS and BE*), the incidence of pathological EDP increased to 76%.

The EDP was raised above 40 mmHg in 47% of the patients, and intermittent increases above 60 mmHg were observed in 30% of the patients.

Table 3. *Ranges of Highest Epidural Pressures (EDP) in 63 Patients After Severe Head Injuries, Related to Simultaneous Findings on CT Scan (group 1, 3, 5) (see* Table 2)

Findings on CT-Scan	Ranges of highest epidural pressures (mm Hg)			
	0–20	21–40	41–60	60
Group 1 N: 10	6/10 (60%)	4/10 (40%)		
Group 3 N: 19	7/19 (37%)	6/19 (31%)	4/19 (21%)	2/19 (11%)
Group 5 N: 34	8/34 (24%)	10/34 (30%)	6/34 (17%)	10/34 (30%)

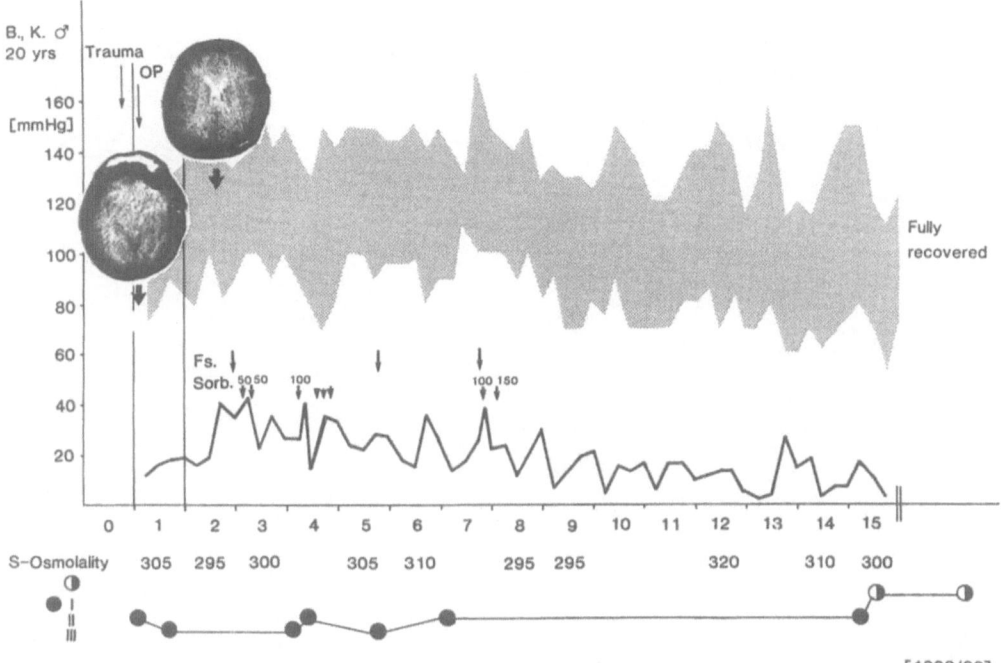

[1333/83]

CT-scan, intracranial pressure and neurologic status in a patient with
epidural hematoma and diffuse brain swelling

almost all cases with marked compression of CSF-spaces. This was particularly marked for the perimesencephalic cistern. Transient pressure increases above 40 mmHg occurred especially in groups 3 and 5.

These findings suggest, that patients with head injuries associated with
general brain swelling, require monitoring of the ICP, if a disturbance of
consciousness is present.

Patients in groups 3 and 5 had a 60% chance of increase in ICP above
40 mm Hg. In patients with obliteration of the intracranial CSF spaces
– the so-called "stiff brain" – mortality was remarkably high, 64%.

Examples 1–3 demonstrate typical courses after severe brain injuries
with development of intracranial space-occupying lesions, such as brain
swelling and/or brain edema.

Example 1 (Fig. 3): A 20-year-old patient with an *early epidural hematoma and diffuse brain swelling (group 3)*. CT of the unconscious patient
some 2 hours after trauma, shows an early epidural hematoma, with signs
of general brain swelling. Following surgical evacuation of the hematoma,
the patient remains unconscious. On day 2, EDP increases up to 40 mmHg
are seen. Under the combined pressure-reducing therapy (hyperventilation,
hyperosmolar boli, and furosemide) EDP shows a slow return to normal at

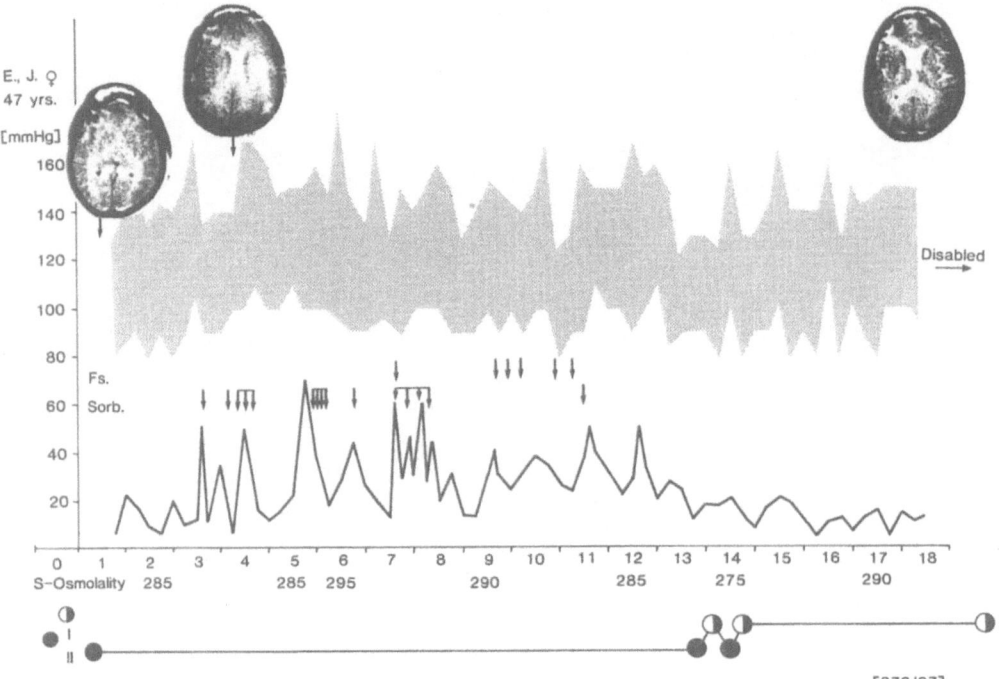

Fig. 4. CT-scan, intracranial pressure and neurologic status in a patient with
diffuse posttraumatic brain swelling and peri-contusional edema

the end of the 2nd week. Simultaneous improvement in the level of consciousness occurred and progressed to full recovery.

Example 2 (Fig. 4): A 47-year-old male patient with *diffuse posttraumatic brain swelling and pericontusional brain edema (group 5)*. The CT-scan some 24 hours after trauma shows brain contusion with early perifocal edema. The patient remains unconscious and on day 3 ICP increases up to 50 mmHg and on day 5 up to 70 mmHg. Under combined pressure reducing therapy, the EDP first returns to normal after day 14.

Following this, there is a continuous improvement in the level of consciousness.

Example 3 (Fig. 5): An 18-year-old female patient with *posttraumatic ischemic brain edema*. After the development of postischemic brain edema on day 6/7 after trauma, simultaneous increase in EDP up to the level of systolic blood pressure is measured. The EDP returns to normal after day 7 under the combined pressure-reducing treatment, however, coma persists, and the patient dies on day 21 after trauma.

These examples clearly illustrate, that with increasing restriction of intracranial compliance due to brain swelling and brain edema ICP increases over a period of 10–14 days after trauma are to be expected.

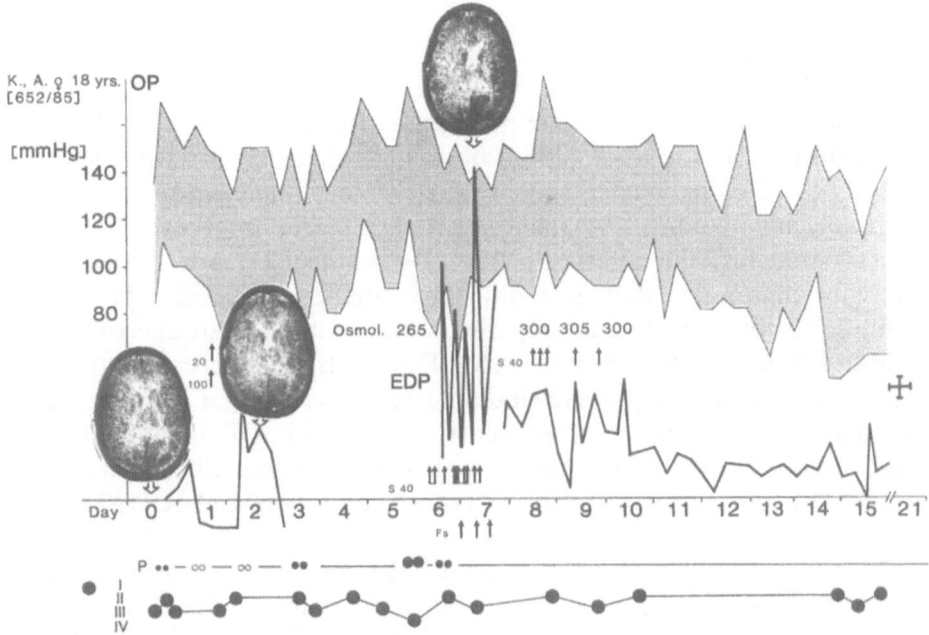

Fig. 5. CT-scan, intracranial pressure and neurologic status in a patient with posttraumatic ischemic brain edema

The development of *posttraumatic ischemic brain edema,* which may lead to fulminating increases in ICP up to the level of cerebral perfusion pressure, *is a* particularly *unfavorable prognostic indicator.* The poor prognosis in this condition serves as a grave reminder for all who treat head-injured patients and the development of therapeutic strategies for its prevention remains an urgent priority.

In a study of 75 comatose patients, Obrist *et al.* (1984) found an ICP increase above 20 mm Hg in 77% of those with hyperemia in contrast to only 23% in those with reduced CBF. Therefore, the most probable explanation for the *relationship between hyperemia and ICP* is an increase in CBV (Langfitt *et al.* 1965, 1983; Obrist *et al.* 1985).

Obrist (1984) and Miller (1987) would expect a CBV increase of 20–30% amounting to as much as 10–15 ml, in the common condition of brain swelling.

When the intracranial compliance is lowered, due to additional space-occupying components, such as extravascular blood or poor CSF- absorption and brain edema, only minor increases in blood volume may cause significant ICP increases (Miller 1975).

Miller (1983) assumed, that BS within the first 24 hours due to an increase in cerebral blood volume was the most frequent cause of early fluctuating rises in ICP in head injured patients. Mendelow *et al.* (1983) however, studied the correlation between CBV and ICP in 52 severely head-injured patients and found no support for this hypothesis.

Murphy, Teasdale *et al.* (1983) have studied the relationship between *ICP and the CT appearance of brain swelling* in patients with head injury. The ventricular volume and the appearances of the 3rd ventricle and the basal cisterns were the criteria noted on 2 separate CT-scans. Patients with diffuse brain injuries had a significantly smaller ventricular volume of 6.1 ml, compared with 16.1 ml in patients with occult hematomas or 29.4 ml in those following evacuation of a clot. Despite these differences, the mean ICP levels of these groups were similar: 19.7–24.5 mmHg. Furthermore, no relationship was found between ventricle size and ICP in patients with diffuse injury. But the authors found a striking association between obliteration of the 3rd ventricle and basal cisterns and increased ICP (Teasdale *et al.* 1986).

Tomei *et al.* (1989) found the trend behaviour of ICP to be the main factor influencing the outcome in a group of 80 patients. Their diagnosis of diffuse brain swelling was based on the reduction or absence of the lateral and third ventricles as well as disappearance of the basal cisterns. No survivors were found in 45% of those patients whose recordings were characterized by monotonous high pulse pressure amplitudes of 50–70 mmHg superimposed on a base line of 20–30 mmHg. The ICP was normal in only 22% of the patients with diffuse brain swelling.

Management of Brain Swelling and Brain Edema: How to Do It?

Analysis of the different time course of brain swelling and brain edema have shown that BS appears early during the first 3 hours after brain injury, with BE appearing at a later stage after the 3rd hour (Fig. 2).

Therefore, the initial priority should be given to means of preventing the posttraumatic brain swelling.

Important *negative factors*, which may enhance the development of BS and BE are summarized in Fig. 6.

These negative factors can be avoided or counteracted if attention is paid to their *general principles of treatment* of head-injured patients, which are listed in Fig. 7.

It is generally believed, that an *optimal fluid balance* is difficult to achieve in patients with BE, because of the risk of aggravating the edema with a resultant increase in ICP. Tramner *et al.* (1989) recently found experimentally that cristalloid infusions aggravated the increase in ICP in

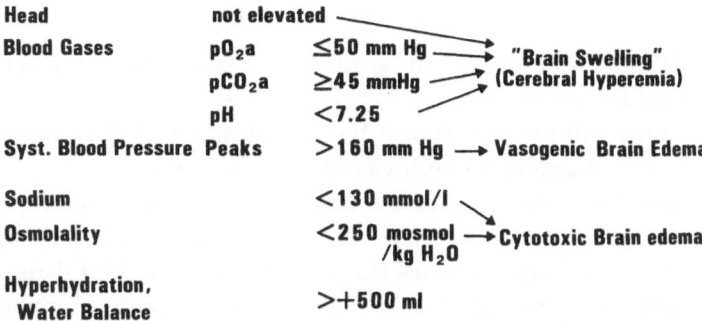

Head	not elevated
Blood Gases	pO_2a ≤ 50 mm Hg ⟶ "Brain Swelling"
	pCO_2a ≥ 45 mmHg (Cerebral Hyperemia)
	pH < 7.25
Syst. Blood Pressure Peaks	> 160 mm Hg ⟶ Vasogenic Brain Edema
Sodium	< 130 mmol/l
Osmolality	< 250 mosmol /kg H_2O ⟶ Cytotoxic Brain edema
Hyperhydration, Water Balance	$> +500$ ml

Fig. 6. Enhancing factors of brain swelling and brain edema

Head	30°–40°
pO_2a	> 70 mm Hg
pCO_2a	25–30 mm Hg
SAPsyst.	100–160mm Hg
Fluid Balance	± 500 ml
Sodium	135–145 mmol/l
Osmolality	270–295 mosmol/kg H_2O

Fig. 7. Basic principles for treatment of head injured patients

dogs with vasogenic BE, whereas colloids did not. Furthermore, colloids improved the neuroelectric status of the edematous brain which was secondary to the increased perfusion. However, in the presence of normal renal function, the infusion of large volumes of various solutions did not affect posttraumatic BE (Shapira *et al.* 1989).

Acute isotonic fluid load did not significantly change brain water content and ICP (James *et al.* 1989).

1. Management of Brain Swelling

1.1. Basic Principles

Basic principles of prevention and treatment of BS are elevation of the head and hyperventilation.

Elevation of the upper part of the body and of the head by 30–40 degrees which supports the cerebral venous return, is an effective and almost always practicable therapeutic measure (Frowein and Richard 1975; Richard 1980/86; Bruce 1982). Extreme elevation of the head, above 40 degree should be avoided as it may lead to a reduction of cerebral blood flow.

1.2. Hyperventilation

Arterial hypoxemia and hypercapnia in comatose head-injured patients are indications for intubation and hyperventilation. Early intubation and hyperventilation will lower cerebral blood volume and ICP, provided that brain vessels still respond to changes in CO_2 tension.

A response to hyperventilation can almost always be expected. As a rule, the CO_2 responsiveness of cerebral vessels is preserved, even in patients with severe brain injuries (Gennarelli et al. 1979; Obrist 1984).

Hyperventilation has the following *beneficial effects* in patients with brain swelling (Reivich 1964, Grubb 1974, Paulsson 1972, Obrist et al. 1985):

- reduction of CBV by cerebral vasoconstriction resulting in a decrease of intracranial volume and in turn a reduction in ICP.
- induction of respiratory alkalosis supports reversal of metabolic acidosis in the brain,
- restoration of CBF autoregulation.

However, both CBF and ICP become refractory to prolonged hyper-ventilation.

Recently Marmarou and Wachi (1989) pointed out, that the blood volume reactivity to PCO_2 change in head-injured surviving patients is relatively constant, and remains intact at levels of mild or aggressive hyperventilation. Only patients who eventually died with a "tight brain", had a reduced blood volume responsivity.

Positive–negative pressure respiration (PNRP) not only causes extreme hypocarbia but also effectively increases flow rate in the superior vena cava (Brecher 1954). It would seem to be an ideal form of treatment for raised ICP, but it has, however, some serious pulmonary side-effects as well as negatively influencing the CBF (Rudenberg et al. 1976).

Therefore, controlled PNPR has to be restricted to a limited period of time, for example during intracranial operative procedures or for lowering life-threatening increases of ICP after trauma.

A slight elevation of *positive end-expiratory pressure* will affect CBV and ICP only in so far that it causes a change in respiratory minute volume or arterial CO_2 tension. Increases in CBV and ICP can be prevented by keeping minute volume and CO_2-tension at a constant level and by elevation of the head (Frost 1977).

It is particularly important for the reduction of ICP, that sufficient *time* is allowed *for expiration*. Otherwise, venous return to the heart may be impaired, producing an increase in central venous pressure (Bruce 1982). In this regard an inspiratory/expiratory ratio of $2:3$ would seem to be especially effective (Richard 1980).

The recent technique of *High Frequency Jet Ventilation* (HFJV) (Merrit 1984; Korn *et al.* 1989; Garcia-Sola *et al.* 1989) lowers both ICP and the global cerebral compliance, possibly more than the conventional mechanical ventilation. This applies, however, only in the early period after trauma and probably not in those patients, who were refractory to pharmacological treatment of elevated ICP.

1.3. Buffering of Brain Tissue Acidosis

Brain tissue acidosis is one of the most important aggravating factors in promoting brain swelling.

In more recent studies, Rosner *et al.* (1989) and Gaab *et al.* (1989) made a case for continuous *thromethamine–*THAM*–*therapy in patients with posttraumatic brain swelling. They were able to measure improvements in the level of CSF lactate, in the ICP and in the duration of the ICP-reducing therapy. They assumed that this was due to an action on the cerebral vasoparalysis.

2. *Management of Traumatic Brain Edema*

The classical distinction between cytotoxic and vasogenic edema (Klatzo 1967) still holds valid, and provided a rationale basis for the development of therapeutic procedures (Hossmann 1986). Following a severe traumatic brain lesion, vasogenic edema develops primarily as a consequence of a massive break down of the BBB. At a later stage, as a result of ischemia, the postischemic type of edema becomes evident. Proton nuclear magnetic resonance (NMR) studies in brain edema have shown, that the maximum extent of edema after vasogenic cold injury, is reached after 24 hours, whereas with triethyl-tin induced cytotoxic edema the maximum is seen after 7 days (Naruse *et al.* 1982)

The *basic principles in the treatment of BE* are prevention or elimination of aggravating factors, such as high peaks of systemic blood pressure, or disturbances of the serum electrolytes and osmolality (Fig. 7).

Therapeutic options may be reviewed with reference to our current knowledge on the evolution and resolution of BE (Bruce 1982; Long 1985),

in short: (1) Removal of mass lesion
        (2) Repair of defective blood brain barrier (BBB)
        (3) Improvement of the microcirculation
        (4) Control of edema mediators
        (5) Support of clearance of edema fluid from the brain

2.1. Removal of Mass Lesion

It has been demonstrated experimentally (Aarabi and Long 1979), that if an area of cortex was removed immediately after being subjected to cold injury, the formation of the expected vasogenic edema was completely abolished. If the lesion was removed 2, 4 or 8 hours later, the advancement of the edema front and an increase in the amount of edema was stopped. *Removing the edema-generating focus, e.g.* area of brain contusion, therefore could be an obvious practical consequence. However, surgical intervention may be of doubtful value in many patients with deep-seated intracerebral contusions, as the avoidance of further damage to potentially viable brain tissue in attempting to reach deep lying vital structures cannot be guaranteed (Bruce 1982; Richard and Frowein 1983).

As it is rarely possible to excise the focus of edema production, therapeutic activity must be concentrated on limiting the spread of edema and on modifying its secondary effects, such as the lowering of tissue pressure and eliminating the influence of extravasated protein (Long 1985).

2.2. Repair of the Defective Blood Brain Barrier (BBB)

Glucocorticoids

Some of the beneficial effects of corticosteroids *e.g.* dexamethasone (dx) has been attributed to its membrane-stabilizing action, *e.g.* capillaries, as well as to the reduction of plasma protein and ionic leakage into the brain (Marshall 1980).

Pappius (1982) found that dx given either before or after the lesion, significantly diminished the depression of local cerebral glucose utilisation seen in untreated animals. This effect was restricted to areas affected by injury. Since the metabolic depression did not parallel the development of cerebral edema, the effects of dx on the functional depression were most probably mediated by the effects on the edema process.

Other authors were not able to confirm, that dx acts by reducing the flow of solutes and water across damaged or leaky vessels in and around the lesion. Experimental findings suggest, that this drug alters the structure of the extracellular space in the tissue around the lesion, and thereby increases the resistance to the flow of the edema fluid from the lesion into the adjacent brain tissue (Yen *et al.* 1985).

Whilst Hartmann *et al.* (1985) found that high-dose dx could prevent the spread of edema caused by cerebral ischemia in infarcted tissue, Facco *et al.* (1985) concluded that dx at a dose of 40 mg/day failed to decrease the progression of perilesional hypodensity and midline shift on CT-scan, in patients with isolated temporal lobe contusions.

The results of treatment in patients with traumatic brain edema using *initial high-dose steroid therapy*, dx, bethametasone, or methylprednisolone, was not comparable to the clinical improvement and reduction in ICP following administration of dx in patients with brain tumors which were associated with perifocal edema (Gudeman *et al.* 1979; Braakmann 1983).

Pitts (1980) and Dearden (1986) observed no change in either the medium pressure or in the average highest peaks of ICP. No difference was noted in the pressure–volume characteristics in their patients.

Cooper *et al.* (1979) found no significant effect on morbidity and mortality in patients following severe head injury treated either with high or low doses of dexamethasone.

Saul *et al.* (1981) obtained similar results, but they were able to distinguish between patients who respond to steroid therapy and non-responders. Their findings were based on assessing the neurological status using a Glasgow Coma Scale rating at certain pre-fixed times. All patients received steroids initially, but therapy was discontinued if no improvement was seen after the 3rd day.

The role of steroid therapy requires further elucidation. Double-blind studies in patients with clearly-defined head injuries are required with the exclusion of those patients who primarily have an especially favorable or unfavorable prognosis (Braakmann 1983). A recent multicenter study, sponsored by Merck LTD, is in progress.

2.3. Improvement of Microcirculation

Disturbances of the cerebrovascular microcirculation are both a cause and a result of BE (Mchedlishvili *et al.* 1986). The improvement of flow in the cerebral microcirculation involves the following factors:

– Blood pressure, blood flow and blood volume in cerebral vessels which are closely related to the ICP
– rheological properties of blood in the cerebral microcirculation.

Klatzo *et al.* (1967) were able to demonstrate the striking effect of systemic hyper- or hypotension on the rate of development of vasogenic BE after cold lesions. Based on these findings, we are confronted with the therapeutic challenge of *maintaining a stable systolic blood pressure between* 100–160 mmHg, depending on the patient's age.

Cerebral capillaries and veins seem to be able to control water efflux, independent of transmural pressure, up to a certain limit. This control, however, is disturbed when brain vessels or brain tissue is damaged.

When this occurs, congestive edema develops (Cuypers *et al.* 1976). When brain tissue is undamaged, venous hypertension alone causes brain swelling.

If however, intracranial venous congestion is sufficiently marked and prolonged, BE was found to occur in all experiments (Mchedlishvili 1986). This observation supports the argument for the maintenance of adequate cerebral venous drainage using hyperventilation therapy.

The driving pressure for vasogenic edema production is the capillary pressure minus the interstitial tissue pressure. Thus, if the end-capillary pressure can be lowered, the production of edema will be decreased. Hyperventilation decreases the end-capillary pressure by arteriolar vasoconstriction (Bruce *et al.* 1982). This beneficial effect of hyperventilation on edema formation will be dependent on the state of CO_2-responsiveness of the cerebral vessels (Obrist *et al.* 1985).

In the ischemic BE after head injury, the inadequacy of blood supply to brain tissue is caused by disturbances in the microcirculation, *e.g.* microthrombi or -emboli, by rheological disturbances, as well as by functional vasoconstriction, the so-called vasospasm. Modern non-invasive measurement of the *cerebral blood flow velocity* (BFV) using ultrasound Doppler sonography, as well as ICP monitoring gives us information about intracranial hemodynamics in patients with posttraumatic brain edema (Sanker *et al.* 1990).

Yoshita *et al.* (1985) studied the influence of the *calcium-antagonist nicardipine* on cortical blood flow and cortical specific gravity following focal cerebral ischemia in cats. They found, as did Steen (1986), that nicardipine, like nimodipine induces a significant alteration in cortical blood flow. Unfortunately it exacerbates the development of early cortical edema through some unknown mechanisms not related to its hemodynamic effect.

Kostron *et al.* (1985) treated 8 patients after head injury, suffering from angiographically confirmed traumatic vasospasms with *nimodipine*. They found a regression of vasospastic narrowing of brain vessels as well as an improvement in the somato-sensory evoked potentials. ICP and mean arterial pressure were not altered.

On the contrary, Guggiari *et al.* (1983) measured a 10% decrease in mean arterial blood pressure and ICP increases of 12% during treatment in 10 patients during the 2nd week following severe head trauma.

A larger randomized double blind study is currently in progress and will possibly clarify these contradictory findings.

Decreasing CBF in postischemic brain edema leads to an *increase of blood viscosity* (Hossmann 1986). For treatment of the postischemic hypoperfusion syndrome, Hossmann (1986) recommends a combination therapy utilising hemodilution, controlled systemic hypertension with dopamine, osmotherapy with 20% sorbitol and anticoagulation with heparin. Other workers found, that the administration of concentrated albumin failed to improve CBF (Little 1981).

2.4. Control of Mediators

The currently available evidence on the role of *mediator compounds*, which could enhance formation and persistance of BE (see page 108) has triggered off the search for *pharmacological inhibitors*. Definitive proof that pharmacological inhibition attenuates brain edema was provided only in the case of the kallikrein–kinine system. Unterberg *et al.* (1987) found, that *aprotinin* was effective in reducing cerebral edema and this was associated with a decrease in tissue water content. However, broad clinical experience with this compound is lacking to date.

At present, investigations on the usefulness of the so-called *free radical scavengers* in the treatment of BE are being undertaken (Asano *et al.* 1985a).

Free radicals have been shown to cause alterations in blood vessel diameter and permeability. Zimmerman (1989) found in animals after focal brain injury, that prophylactically given free radical scavengers, such as superoxide dismutase (SOD) and catalase, did not prevent cerebral vascular congestion, but did cause a significant reduction of the progression of intracranial hypertension, possibly by reducing edema formation.

2.5. Support of the Edema Clearance Rate

Another approach to the treatment of cerebral edema is to *increase* the *edema clearance rate* from the brain tissue. This is based on the concept, that entry of vasogenic edema fluid into the CSF-space (*e.g.* the ventricles) is one of the main mechanisms of resolution of vasogenic BE (Reulen *et al.* 1978).

The *lowering* of the *ventricular fluid pressure* (VFP) by steroids, diuretics or by external CSF drainage is a logical therapeutical step to support the removal of intra-parenchymal brain water associated with brain edema. Marmarou *et al.* (1986), however, recorded a constant net CSF-formation of approximately 500 ml/day over the first 5 days after trauma in patients with severe head injuries. They concluded, that ventricular clearance of edema is minimal during this period in the majority of patients. Whether the lowering of VFP may result in a more rapid clearance of edema fluid, remains to date an unsolved question (Bruce 1982).

Moreover, it is often difficult to obtain a satisfactory decrease of VFP using external CSF drainage, because the lateral ventricles are reduced in size during the first days after injury (Miller 1983)

3. Treatment of Severe ICP Increases in Traumatic Brain Edema

The use of osmotic diuretics, loop diuretics and barbiturates are further therapeutic means of treating BE, but should only be used under continuous ICP monitoring (Richard 1980).

3.1. Hyperosmotic Solutions

The most frequently used osmotically effective agents are mannitol, sorbitol and glycerol.

Despite the fact that hypertonic solutions have been used since the early part of this century, there is still disagreement on their *mechanisms of action* (Langfitt 1986).

Muizelaar *et al.* (1984) have presented evidence, that the mechanism of autoregulatory vasoconstriction may just be as important as dehydration of the brain in reducing the ICP.

However, Auer and Haselsberger (1986) argue that vasoconstriction plays only a secondary role in the lowering of ICP after mannitol therapy. They observed only modest changes in the caliber of the pial arteries and veins and doubted that these could be responsible for the reduction in ICP.

Albright *et al.* (1984) proposed that the main effect was an osmotic reduction of brain water content, as long as the blood brain barrier was intact. Nath and Galbraith (1986) were able to measure an increase in brain specific gravity after *mannitol* infusion in 13 patients with an intracranial hematoma. This would confirm, that administration of mannitol reduces the water content of the surrounding brain tissue.

Sorbitol in a 40% solution is especially effective and has a very high osmotic pressure of 3200 mmosmol/kg H_2O. A rapid infusion of sorbitol causes an immediate steep increase in serum osmolality and leads in turn to a prompt decrease in ICP, which may even reach subatmospheric levels (Richard 1980).

Since edema is not to be expected before the 3rd hour after a severe brain injury, osmotic diuretics are usually not indicated before arrival in hospital (Richard and Karimi 1986). Small volumes of sorbitol given as a *bolus* were effective in lowering ICP in pericontusional or in postischemic BE. However, repeated application of hyperosmotic solutions should be avoided unless regular measurement of the serum osmolality is carried out. Otherwise, renal failure may be precipitated with secondary enhancement of the BE. In addition, the serum lactate levels must be monitored in patients receiving sorbitol treatment. Furthermore, acute hypervolemia with increased central venous pressure can lead to complications such as arrhythmias, particularly in patients with congestive heart failure (Richard 1986).

The effect of *glycerol* in BE and elevated ICP, first reported by Cantore *et al.* (1964), is based on osmotic and metabolic actions and is effective both via the oral and the intravenous route (Csandra *et al.* 1985).

Using positron emission tomography, Masatsuma *et al.* (1988) were able to measure increases in CBF in intact non-edematous brain tissue, which led to an improvement of the functional reserve for oxygen metabolism.

Compared to mannitol, glycerol seems to cause a smaller decrease in ICP, no rebound phenomenon is observed, and the decrease in CBF is less abrupt and of longer duration (Garcia-Sola *et al.* 1989).

Hartmann recently found, that increases in local CBF after administration of mannitol, sorbitol or glycerol did not parallel alterations of rheologic parameters and changes of osmolality. Only mannitol led regularly to long-term rebound phenomena (Garcia-Sola 1989).

3.2. Loop Diuretics

Simultaneous application of diuretics, such as *furosemide*, causes a forced diuresis and increases the ICP-lowering potential of hyperosmotic solutions. In addition, diuretics have a direct effect at membrane level in preventing absorption of water into glial tissue (Miller 1983).

Loop-diuretics prolong the bulk movement of water and ions across the BBB, due primarily to their renal effect (Schettini *et al.* 1982).

It is essential that the development of electrolyte disturbances, hypovolemia and hypo- or hyperosmolality should be avoided at all costs.

3.3. Barbiturates

Barbiturates are presumed to reduce cerebral metabolism and ICP by means of cerebral vasoconstriction. Some investigators maintain that barbiturates can control ICP in severely head injured patients, when hyperventilation and osmotic solutions have failed (Marshall *et al.* 1979). This treatment, however, is not without *risk* (Miller 1987), as a serious potential danger is the development of arterial hypotension, especially in the hypovolemic patient.

The short acting anesthetic agent *gammahydroxybutyrate* has also been used for this purpose and produces a fall in blood pressure (Leggate *et al.* 1986). Continuous monitoring of systemic blood pressure, as well as ICP and central venous pressure, is essential in each patient to ensure maintenance or improvement of cerebral perfusion pressure.

Recent studies assessing the action of barbiturates on the outcome in head injured patients, showed little or no beneficial effects as they failed to improve the survival rate or the quality of survival (Ward *et al.* 1986).

Early Outcome and Follow-up After Posttraumatic Brain Swelling and Brain Edema

1. Early Outcome

In our patient group (see page 112) computed tomography did not reveal any pathological alterations in 42 out of 273 severely head injured patients. The posttraumatic mortality of this group was 5% (Table 2).

In patients with diffuse brain swelling without any additional lesions, the mortality increased slightly up to 9% and reached 20% in patients with contusions or hematomas without signs of brain swelling or brain edema.

The highest mortality (47%) was observed in patients with contusional hemorrhages or hematomas with brain swelling.

Mortality in the patient group with intracranial bleeding and signs of brain swelling and brain edema, was 28% (group 5). The higher mortality in group 3 did not correlate with the increasing age of the patients (see Table 2).

Probably the most important reason for the *high mortality in patients with intracranial bleeding and signs of brain swelling* (group 3), is the extremely rapid exhaustion of intracranial space reserve. *68%* of these patients *died within 48 hours* after trauma, as opposed to only 13% in group 5 over the same period. Therefore, the survival time of the patients in group 3 was too short as a rule for the development of recognizable signs of brain edema on CT.

This analysis of the traumatic sequelae in the CT scans of the head reveals that brain swelling and brain edema are processes with a distinctly different time course. After severe brain injury, brain swelling develops very frequently within the first 3 hours, both in children and juveniles, but also in elderly patients. Brain swelling leads very rapidly to a life-threatening restriction of intracranial compliance, particularly when associated with a hematoma or a contusion.

Brain edema develops at a later stage between the 3rd and the 6th hour at the earliest. An even higher incidence is found after the 12th hour, and is seen quite often as an additional complication of therapy-resistant brain swelling (Fig. 2).

We found a good correlation between ICP, grades of brain swelling in CT, and patient outcome on 36 cases with contusional hemorrhages (Richard *et al.* 1986). Furthermore, the extent of pericontusional edema was found to be related to the prognosis (Table 4).

2. Follow-up

The study of Lobato *et al.* (1983) was based on the analysis of the influence of the type of intracranial lesion on the final outcome using serial CT scanning. This study of a consecutive series of 277 patients with severe head injury revealed a poor outcome in patients with acute hemispheric swelling after operative removal of a large extracerebral hematoma. All these patients died. By way of contrast, in the same study only 4 out of 41 patients died (10%) with generalized brain swelling. The authors assert that the reduction of CSF-spaces leads to a decrease in intracranial compliance,

Table 4. *Outcome of Patients with Brain Contusions in Relation to Volumes of Contusional Hemorrhages (CH), Edematous Reaction (ER), Grades of Brain Swelling and Coma, and ICP Maxima (from Richard et al. 1986)*

Outcome	N	%	Age [x̄]	CH-Volume [ml]	ER-Volume [ml]	Brain Swelling [Grade]	Coma [Grade]	ICP mx [mm Hg]
Full recovery	4	11	43	20	59	2	1.8	26
Moderate disability	5	14	39	8	18	2.4	2.3	35
Severe disability	10	28	15	3	10	2.9	2.8	52
Lethal outcome	17	47	33	16	30	2.9	2.7	77
Σ	36	100	29					

Table 5. *Early Outcome (O) and Follow-up (F) of Patients After Severe Head Injury in Relation to Age and Posttraumatic CT-findings*

Outcome symbols (columns for each age group): ↑ | (↑) | (↓) | ↓ | + | [+]

GROUPS	CT-Findings	O/F	No. of Patients	Age x̄	SD	All Age Groups (O 273 / F 269) ↑	(↑)	(↓)	↓	+	[+]	Child + Juveniles (O 54 / F 53) ↑	(↑)	(↓)	↓	+	[+]	Adults (O 98 / F 95) ↑	(↑)	(↓)	↓	+	[+]	Middle Aged (O 71 / F 71) ↑	(↑)	(↓)	↓	+	[+]	Seniors (O 50 / F 50) ↑	(↑)	(↓)	↓	+	[+]
5	Contusion/Hematoma + Brain Swelling + Brain Edema	O	85	43	21	14			47		24	1			5			6			13		19		15	9	29	15	8	7	4	2	19	2	3
5		F	85	43	21	17	13	10	9	24	12	5	2	1		3	3	8	3	1	4	7	3	4	6	7	2	7	5	2	1			7	6
4	Contusion/Hematoma + Brain Edema	O	5	54	19	3					2							1						2	1										1
4		F	5	54	19	1	2	1			1	1					1	1			1		1	2	1							1			1
3	Contusion/Hematoma + Brain Swelling	O	88	39	21	23			41		24	5					6	7			9		19	8	3		6	4	5	3		2	6	10	3
3		F	86	39	21	22	8	3	5	41	7	8	1				6	7	4	3	1			5		2	6	4	2		2	10	3
2	Contusion/Hematoma	O	10	59	24	3			5		2	1						1			1					2			1		2			2
2		F	10	59	24	2	2	1	1	2	2	1					1	1	1		1			1	1	1			1		2			2
1	Brain Swelling	O	43	29	18	19			20		4	5			7		1	8			9		1	4	2		4	1	1	2			1		
1		F	41	29	18	27	8	1	1	4		10	2				1	11	3	1		1	1	5	2		1	1		1			1		
0		O	42	37	22	31			9		2	7			3			14			2			6	1		1	1	1	4	3		1		
0		F	42	37	22	29	5	2	2	2	2	7	1	1	1	1	1	14	2			1	1	5	1	1		1	1	3	1		1	1	1

↑: fully recovered; (↑): slightly disabled; (↓): severely disabled but independent of care; ↓: in need of care; +: deceased during follow-up. [+]: deceased during hospitalisation.

allowing little room for expansion for associated focal lesions. They observed that an elevated ICP returns to normal following evacuation of a small epidural hematoma.

Until now, little is known about the follow-up of patients with traumatic swelling and/or brain edema in the long run. Therefore, we reassessed the follow-up of our patients in 1989, 3–4 years after the injury (Table 5).

On 4 patients we were not able to get any information. 24 patients died posterior to the discharge from our hospital. Out of these, 12 belonged to the CT-group 5. Six of these patients were seniors. 13 out of the 24 patients, deceased in the follow-up, were older than 60 years. None of the young patients died in the later course, but 3 out of the adults with a group-5-brain injury and 8 of the middle-aged with group-3/or -5-brain injuries. Follow-up data reveal that surviving patients up to 40 years of age have had a significant chance for full recovery. The number of the young patients with complete recovery increased from 19 to 31, i.e. 63%, the number of adult patients from 37 to 41, i.e. 10%. Contrarily, the number of older patients who recovered completely, decreased in the follow-up by 29% for the middle-aged and by 33% for seniors. Patients with brain swelling but without additional lesions had the highest growth rate in full recoveries in the long run: 42%. This figure was even better for young patients: 100%.

The incidences (n) of these various combinations of contusions, hematomas, brain swelling and brain edema in our 273 consecutive patients, the frequency (%) of full recovery and of fatal courses in the early outcome (O) and in the follow-up (F) for each group, are calculated in Table 5, and represented in Fig. 8.

Contusion without other lesions, or contusion with brain edema only on the initial posttraumatic CT were rare entities. The increasing mortality of the older patients in the later follow-up was due to non-traumatic courses in half of the cases.

No visible lesion in the first CT often resulted in full recovery in younger patients but was associated with increasing risk of mortality for older adults and seniors.

Isolated brain swelling was more frequent in children, juveniles and younger adults; it resulted in about half of the cases in early full recovery and on even higher percentage at the later follow-up. There was low mortality.

The combination of contusion and/or hematomas with early brain swelling clearly reduced the chance of full recovery, predominantly in children and juveniles; the mortality already reached up to 50% and more. In children and juveniles the number of patients with full recovery increased significantly in the later follow-up.

With the combination of contusions plus swelling plus edema, full recovery was less likely. No older patient recovered completely. The risk of

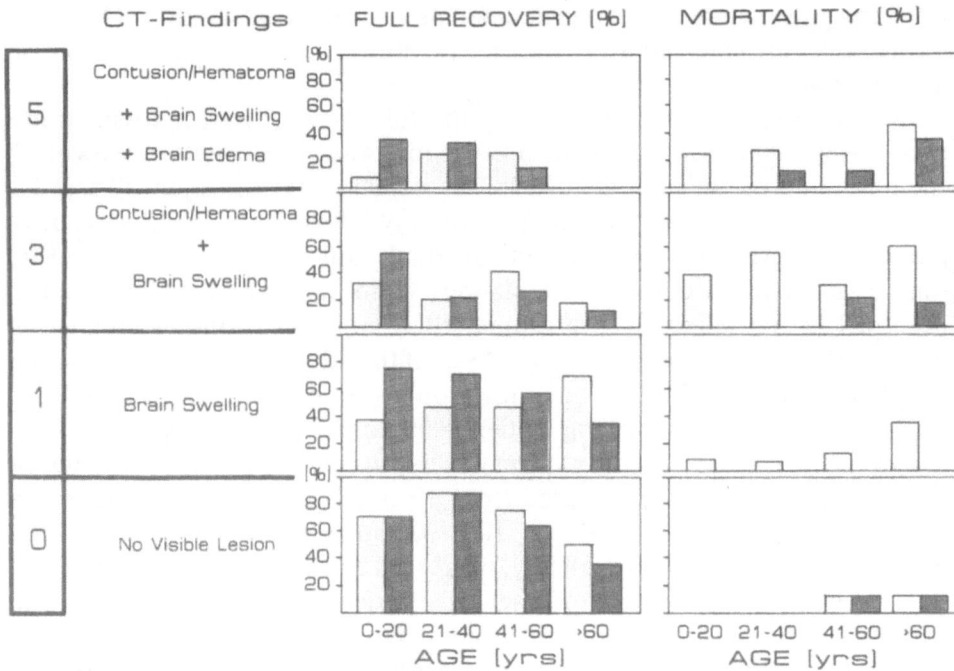

Fig. 8. Percentages of full recovery and mortality in the early outcome (first columns) and the follow-up (second columns) in relation to patient's age and CT-findings

a fatal issue was about 20–30%, even higher up to 80% in the senior patients, summarizing early and later mortality.

This implies, that brain swelling, contusions, or hematoma alone appear to be mostly treatable lesions. The main risk for the patients arises especially from the rapid exhaustion of intracranial space capacity by brain swelling in addition to space-occupying lesions such as contusions and hematomas.

Also the development of posttraumatic ischemic brain edema, which may lead to fulminating increases in ICP up to the level of cerebral perfusion pressure, is an unfavorable prognostic indicator. The poor prognosis in these conditions serve as a grave reminder and the development of therapeutic strategies for their prevention remains an urgent problem.

References

1. Aarabi B, Long DM (1979) Dynamics of an intact vascular bed in the production and propagation of vasogenic brain edema. J Neurosurg 51: 779–784

2. Albright AL, Latschaw RE, Robinson AG (1984) Intracranial and systemic effects of osmotic and oncotic therapy in experimental brain edema. J Neurosurg 60: 481–489

3. Asano T, Gotoh O, Koida T, Takakura K (1985) Ischemic brain edema following occlusion of the middle cerebral artery in the rat. II Alteration of eicosanoid synthesis profile of brain microvessels. Stroke 16: 110–113

4. Auer LM, Haselsberger K (1986) The effect of intravenous manitol on cat pial arteries and veins during normal and elevated intracranial pressure. In: Miller JD *et al* (eds) Intracranial pressure VI. Springer, Berlin Heidelberg New York Tokyo, pp 585–589

5. Baethmann A (1978) Pathophysiological and pathochemical aspects of cerebral edema. Neurosurg Rev 1: 85–100

6. Baethmann A, Maier-Hauff K, Schürer L, Lange M, Kempski O, Unterberg A (1983) The effect of high intracranial pressure on activation and release of mediator compounds in traumatic vasogenic brain edema. In: Ishii S *et al* (eds) Intracranial pressure V. Springer, Berlin Heidelberg New York, pp 405–412

7. Becker DP (1986) The temporal genesis of primary and secondary brain damage in experimental and clinical head injury. In: Baethmann A *et al* (eds) Mechanisms of secondary brain damage. Plenum Press, New York, pp 47–64

8. Braakman R, Schouten HJA, Blaauw-van Dishoeck M, Minderhoud JM (1983) Megadose steroids in severe head injury. Results of a prospective double-blind clinical trial. J Neurosurg 58: 326–330

9. Brecher GA (1956) Venous return. Grune and Stratton, London

10. Bruce DA, Alavi A, Bilianuk L, Dolinskas C, Obrist W, Uzzell B (1981) Diffuse cerebral swelling following head injuries in children: the syndrome of "malignant brain edema" J Neurosurg 54: 170–178

11. Bruce DA, Sutton LN, Schut L (1982) Acute brain swelling and cerebral edema in children. In: De Vlieger M *et al* (eds) Brain edema. J Wiley and Sons, New York, pp. 125–145

12. Bullock R, Blake G, di Trevu M, Favier J (1986) The value of CT scan density measurements after human head injury: A comparative study using microgravimetric measurement of brain specific gravity. In: Miller JD *et al* (eds) Intracranial pressure VI. Springer, Berlin Heidelberg New York Tokyo, pp 20–24

13. Cantore G, Guidetti B, Virno M (1964) Oral glycerol for the reduction of intracranial pressure. J Neurosurg 21: 278–290

14. Clasen RA, Guariglia RJ, Stein RJ, Pandolfi S, Lobiek JJ (1986) Histopathology and computerized tomography of human traumatic cerebral swelling. In: Baethmann A *et al* (eds) Mechanisms of secondary brain damage. Plenum Press, New York, pp 29–45

15. Cooper P, Moody S, Clark WK (1979) Dexamethasone and severe head injury. A prospective double-blind study. J Neurosurg 51: 307–316

16. Csanda E, Basky F, Tulok J (1985) Effectiveness and mode of action of glycerol in brain edema therapy. In: Inaba Y *et al* (eds) Brain edema. Springer, Berlin Heidelberg New York, pp 546–549

17. Cuypers J, Matakas F, Potolicchio SJ (1976) Effects of cerebral venous pressure on brain tissue pressure and brain volume. J Neurosurg 45: 89–94
18. Dearden NN, Gibson JS, McDowall DG (1986) Effect of high dose dexamethasone on outcome from severe head injury. J Neurosurg 64: 81–88
19. Enevoldsen EM, Cold G, Jensen FT, Malmros R (1976) Dynamic changes in regional CBF, intraventricular pressure, CSF pH and lactate levels during the acute phase of head injury. J Neurosurg 44: 191–214
20. Facco F, Zuccarello M, Andrioli GC (1985) Dexamethasone fails to prevent the development of brain edema in temporal lobe contusions In: Inaba Y *et al* (eds) Brain edema. Springer, Berlin Heidelberg New York, pp 528–532
21. Fishman RA (1955) Brain edema. New Engl J Med 293: 706–711
22. Frost EAM (1977) Effects of positive endexpiratory pressure on ICP and compliance in brain injured patients. J Neurosurg 47: 195–200
23. Frowein RA, Richard KE (1975) Allgemeine pathophysiologische Grundlagen zerebraler Schädigungen. In: Lindenschmidt TO (ed) Pathophysiologische Grundlagen der Chirurgie. Thieme, Stuttgart, pp 697–761
24. Gaab MR, Seegers K, Goetz C, Smedena RJ (1989) THAM (Thromethamine): Effective therapy of traumatic brain swelling? In: Hoff JT, Betz AL (eds) Intracranial pressure VII. Springer, Berlin Heidelberg New York, pp 616–619
25. Garcia-Sola R, Rubio JJ, Gilsanz F (1989a) High frequency ventilation versus conventional mechanical ventilation. Their influence on cerebral elastance. In: Hoff JT, Betz AL (Eds) Intracranial pressure VII. Springer, Berlin Heidelberg New York, pp 275–277
26. Garcia-Sola R, Gilsanz F, y Chillon D (1989b) Immediate and long-term effects of mannitol and glycerol. Comparative experimental study. In: Hoff JT, Betz AL (eds) Intracranial pressure VI. Springer, Berlin Heidelberg New York, pp 451–453
27. Gennarelli TA, Obrist WD, Langfitt TW (1979) Vascular and metabolic reactivity to changes in PCO_2 in head injured patients. In: Popp AS *et al* (eds) Neural trauma. Raven Press, New York, pp 1–7
28. Grubb RL, Raichle ME, Eichling JO (1974) The effects of changes in $PaCO_2$ on cerebral blood volume, blood flow, and vascular mean transit time. Stroke 5: 630–639
29. Gudeman SK, Miller JD, Becker DP (1979) Failure of high dose steroid therapy to influence ICP in patients with severe head injury. J Neurosurg 51: 301–306
30. Guggiari M, Guillaume A, Dagreou F (1983) Intracranial pressure (ICP) and hemodynamic effects of a new calcium blocking agent: Nimodipine. Anesthesiology 59: A 357
31. Hartmann A, Hossmann KA, Czernicki Z (1985) Effect of dexamethasone on regional CBF and on serum protein extravasation in experimental brain infarcts in monkeys. In: Inaba Y *et al* (eds) Brain edema. Springer, Berlin Heidelberg New York, pp 646–651
32. Hossmann KA (1986) The role of recirculation for functional and metabolic recovery after cerebral ischemia. In: Baethmann A *et al* (eds) Mechanisms of

secondary brain damage. Plenum Press, New York, pp 239–248

33. Hossmann KA, Matsucka Y (1983) Influence of tissue osmolality on intracellular fluid shifts and the development of ischemic brain edema. In: Reivich et al (eds) Cerebrovascular diseases. Raven Press, New York, pp 183–192

34. Houthoff HJ (1987) Pathobiology of blood brain barrier and brain edema. In: Cohadon F et al (eds) Traumatic brain edema. Liviana Press, Padova, pp 1–14

35. Ishikawa M, Kikuchi H, Nishizawa H, Kobayashi A (1989) Effects of glycerol on CBF and oxygen metabolism. In: Hoff JT, Betz AL (eds) Intracranial pressure VII. Springer, Berlin Heidelberg New York, pp 907–910

36. Ito, U, Seida M, Tomida S, Yamazaki S, Inaba Y (1985) Acute brain swelling, contusional brain edema, and ischemic brain damage in acute head injury. In: Inaba Y et al (eds) Brain Edema. Springer, Berlin Heidelberg New York, pp 621–631

37. James HE, Schneider S (1989) Experimental cerebral edema, isotonic intravenous infusions, mannitol, serum osmolality, electrolytes, brain water and ICP. In: Hoff JT, Betz AL (eds) Intracranial pressure VII. Springer, Berlin Heidelberg New York, pp 459–462

38. Kempski O (1986) The cell swelling mechanism in brain. In: Baethmann A et al (eds) Mechanisms of secondary brain damage. Plenum Press, New York, pp 203–220

39. Klatzo, I (1967) Neuropathological aspects of brain edema. J Neuropathol Exp Neurol 26: 1–14

40. Klatzo I, Wisniewski H, Steinwall O, Streichner E (1967) Dynamics of cold injury edema. In: Klatzo I et al (eds) Brain edema. Springer, New York, pp 554–563

41. Korn A, Aloy A, Czeck Th (1989) The effect of superimposed high-frequency auxiliary ventilation on intracranial pressure. In: Hoff JT, Betz AL (eds) Intracranial pressure VII. Springer, Berlin Heidelberg New York, pp 497–501

42. Kostron H, Rumpl E, Stampfl G (1985) Treatment of cerebral vasospasm following severe head injury with the calcium influx blocker nimodipine. Neurochirurgia 28: 103–109

43. Kuhl DE, Alavi A, Hoffmann EJ (1980) Local cerebral blood volume in head-injured patients. J Neurosurg 52: 309–320

44. Langfitt TW (1983) CT, NMR, and emission tomography in the diagnosis and management of brain swelling and intracranial hypertension. In: Ishii S et al (eds) Intracranial pressure V. Springer, Berlin Heidelberg New York, pp 54–67

45. Langfitt TW (1986) Raised ICP in head injury, its significance, causes and therapy. In: Miller JD et al (eds) Intracranial pressure VI. Springer, Berlin Heidelberg New York, pp 789–794

46. Langfitt TW, Tannanbaum HM, Kassell NF (1966) The etiology of acute brain swelling following experimental head injury. J Neurosurg 24: 47–56

47. Lanksch W, Baethmann A, Kazner E (1982) Computed tomography of brain edema. In: De Vlieger M et al (eds) Brain edema. John Wiley and Sons, New York, pp 67–115

48. Leggate JRS, Dearden NM, Miller JD (1986) The effects of gammahyd-roxybutyrate and sodium thiopentone on ICP in severe head injury. In: Miller JD *et al* (eds) Intracranial pressure VI. Springer, Berlin Heidelberg New York, pp 754–756

49. Lewelt W, Jenkins LW, Miller JD (1980) Autoregulation of cerebral blood flow after experimental fluid percussion injury of the brain. J Neurosurg 53: 500–511

50. Little JR, Slugg RM, Latchaw JP (1981) Treatment of acute focal cerebral ischemia with concentrated albumin. Neurosurgery 9: 552–558

51. Lobato RD, Cordobes F, Rivas JJ (1983) Outcome from severe head injury related to type of intracranial lesion. A computerized tomography study. J Neurosurg 59: 762–774

52. Long DM (1985) New therapies for brain edema. In: Inaba Y *et al* (eds) Brain edema. Springer, Berlin Heidelberg New York, pp 565–577

53. Marmarou A, Maset AL, Ward JD (1986) Dynamics of intracranial pressure rise in severely head injured patients. In: Miller JD *et al* (eds) Intracranial pressure VI. Springer, Berlin Heidelberg New York, pp 9–14

54. Marmarou A, Wachi A (1989) Blood volume responsitivity to ICP change in head injured patients. In: Hof JT, Betz AL (eds) Intracranial pressure VII. Springer, Berlin Heidelberg New York, pp 688–690

55. Marshall LF (1980) Treatment of brain swelling and brain edema in man. Adv Neurol 28: 459–469

56. Marshall LF, Smith RW, Shapiro HM (1979) The outcome with aggressive treatment in severe head injuries. II. Acute and chronic barbiturate administration in the management of head injury. J Neurosurg 50: 26–30

57. Mchedlishvili G, Cervos-Navarro J, Hossmann KA, Klatzo I (eds) (1986) Brain edema. A pathogenetic analysis. Akademiai Kiado, Budapest

58. Mendelow AD, Teasdale G, Teasdale E, Matheson M, Russell T (1983) Cerebral blood volume and intracranial pressure in head injured patients. In: Ishii S *et al* (eds) Intracranial pressure V. Springer, Berlin Heidelberg New York, pp 495–500

59. Merrit J (1984) The effect of high frequency jet ventilation on ICP in patients with closed head injuries. J Trauma 24: 73–75

60. Miller JD (1975) Volume and pressure in the craniospinal axis. Clin Neurosurg 22: 76–105

61. Miller JD (1983) Significance and management of intracranial hypertension in head injury. In: Ishii S *et al* (eds) Intracranial pressure V. Springer, Berlin Heidelberg New York, pp 44–53

62. Miller JD (1987) Brain edema in human head injury. In: Cohadon F *et al* (eds) Traumatic brain edema. Springer, Berlin Heidelberg New York, pp 99–104

63. Miller JD, Corales RL (1982) Brain edema as a result of head injury: Fact or fallacy. In: De Vlieger M, de Lange SA, Beks JWF (eds) Brain edema. John Wiley and Sons, New York, pp 99–114

64. Muizelaar JP, Lutz HA, Becker DP (1984) Effect of mannitol on ICP and

CBF and correlation with pressure autoregulation in severely head-injured patients. J Neurosurg 61: 700–706

65. Murphy A, Teasdale E, Matheson M (1983) Relationship between CT indices of brain swelling and intracranial pressure after head injury. In: Ishii S *et al* (eds) Intracranial pressure V. Springer, Berlin Heidelberg New York, pp 562–566

66. Naht F, Galbraith S (1986) The effect of mannitol on water content of white matter after head injury in man. In: Miller JD *et al* (eds) Intracranial pressure VI. Springer, Berlin Heidelberg New York, pp 581–583

67. Naruse S, Horikawa Y, Tanaka Ch (1982) Proton nuclear magnetic resonance studies on brain edema. J Neurosurg 56: 747–752

68. Obrist WD, Langfitt TW, Jaggi JL (1984) Cerebral blood flow and metabolism in comatose patients with acute head injury. Relationship to intracranial hypertension. J Neurosurg 61: 241–253

69. Ommaya AK, Rockhoff DD, Baldwin M (1964) Experimental concussion. J Neurosurg 21: 249–265

70. Pappius HM (1982) Dexamethasone and local cerebral glucose utilisation in freeze traumatized rat brain. Ann Neurol 12: 157–162

71. Paulson OB, Olesen J, Christensen MS (1972) Restoration of autoregulation of cerebral blood flow by hypocapnia. Neurology 22: 286–293

72. Pitts LH, Kaktis JV (1980) Effect of megadose steroids on ICP in traumatic coma. In: Shulman K *et al* (eds) ICP IV. Springer, Berlin Heidelberg New York, pp 638–642

73. Povlishok JT, Kontos HA (1982) The pathophysiology of pial and intraparenchymal vascular dysfunction. In: Grossmann RS *et al* (eds) Head injury, basic and clinical aspects. Raven Press, New York

74. Reichardt M (1905) Zur Entstehung des Hirndrucks bei Hirngeschwülsten und anderen Hirnkrankheiten und die bei diesen zu beobachtende besondere Form der Hirnschwellung. Dtsch Z Nervenheilk 28: 306–361

75. Reivich M (1964) Arterial PCO_2 and cerebral hemodynamics. Am J Physiol 206: 25–31

76. Reulen HJ (1987) Brain edema in human head injury. Round table. In: Cohadon F *et al* (eds) Traumatic brain edema. Springer, Berlin Heidelberg New York, pp 112–113

77. Reulen HJ, Tsuyumi M, Tack A (1978) Clearance of edema fluid into cerebrospinal fluid. J Neurosurg 48: 754–764

78. Reulen HJ, Tsuyumi M (1982) Pathophysiology of formation and natural resolution of vasogenic brain edema. In: De Vlieger *et al* (eds) Brain edema. J Wiley and Sons, New York, pp 31–48

79. Richard KE (1980) Intrakranielle Drucksteigerung, ihre Pathogenese, Klinik und Behandlung. Nervenarzt 51: 392–405

80. Richard KE, Frowein RA (1980) Prognostic significance of intracranial pressure and neurological condition in acute brain lesions. In: Shulman K *et al* (eds) Intracranial pressure IV. Springer, Berlin Heidelberg New York, pp 10–16

81. Richard KE, Frowein KE (1983) The value of ICP-monitoring in the treat-
    ment of traumatic bilaterally or medially situated intracerebral contusional
    hemorrhages. In: Ishii S *et al* (eds) Intracranial pressure V. Springer, Berlin
    Heidelberg New York, pp 517–526
82. Richard KE, Karimi A (1986) Treatment of intracranial hypertension without
    barbiturates. Advances in Neurosurgery 14: 317–323
83. Richard KE, Radebold K, Frowein RA (1986) Contusional hemorrhage.
    Prognostic significance of primary and secondary brain damage. In:
    Baethmann A *et al* (eds) Mechanisms of secondary brain damage. Plenum
    Press, New York, pp 341–348
84. Richard KE, Wirtelarz R, Frowein RA (1989) Frequency and prognosis of
    traumatic brain edema. Advances in Neurosurgery 17: 81–86
85. Rosner MJ, Elias KG, Coley J (1989) Prospective, randomized trial of THAM
    therapy in severe brain injury: preliminary results. In: Hoff JT, Betz AL (eds)
    Intracranial pressure VII. Springer, Berlin Heidelberg New York, pp
    611–615
86. Rudenberg FR, Mc Graw CP, Tindall GT (1976) Effects of hyperventilation,
    CO_2, and CSF pressure on internal carotid blood flow in baboon. J Neu-
    rosurg 44: 347–352
87. Sakamoto H, Tanaka K, Nakamura T (1986) Direct observation of auton-
    omic nerve activity in experimental acute brain swelling. In: Miller JD *et al*
    (eds) Intracranial pressure VI. Springer, Berlin Heidelberg New York, pp
    166–173
88. Sanker P, Terhaag D, Richard KE, Frowein RA (1990) Cerebrale
    Blutflußgeschwindigkeit: ein prognostischer Faktor nach schweren Schädel-
    hirntraumen? Acta traumatol. 20: 152–156
89. Saul TG, Ducker TB, Salcman M, Carro E (1981) Steroids in severe head
    injury: A prospective randomized clinical study. J Neurosurg 54: 596–600
90. Schettini A, Stahurski B, Young HF (1982) Osmotic and oncotic loop diuresis
    in brain surgery. J Neurosurg 56: 679–684
91. Shapira Y, Muggia-Sullam M, Freund HR, Cotev S (1989) The effect of
    intravenous fluids on cerebral edema after experimental blunt head trauma.
    In: Hoff JT, Betz AL (eds) Intracranial pressure VII. Springer, Berlin Heidel-
    berg New York, pp 995–997
92. Steen PA, Gisvold SE, Milde JH (1985) Nimodipine improves outcome when
    given after complete cerebral ischemia in primates. Anesthesiology 62:
    406–414
93. Teasdale GM, Mendelow AD, Galbraith S (1986) Causes and consequences of
    raised intracranial pressure in head injuries. In: Miller JD *et al* (eds) Intra-
    cranial pressure VI. Springer, Berlin Heidelberg New York, pp 3–8
94. Tomei G, Sganzela E, Spagnoli D (1989) Relationship between clinical course,
    CT scan and ICP in posttraumatic diffuse lesions. In: Hoff JT, Betz AL (eds)
    Intracranial pressure VII. Springer, Berlin Heidelberg New York, pp
    625–629
95. Tornheim PA, Mc Laurin RL (1981) Acute changes in regional brain water

content following experimental closed head injury. J Neurosurg 55: 407–413

96. Tramner B, Iacobacci R, Kindt G (1989) Colloid volume expansion and cerebral brain edema. In: Hoff JT, Betz AL (eds) Intracranial pressure VII. Springer, Berlin Heidelberg New York, pp 992–994

97. Unterberg A (1987) Mediators of brain edema. Mechanisms, formation or release and therapeutic inhibition. In: Cohadon F *et al* (eds) Traumatic brain edema. Liviana Press, Padova, pp 135–142

98. Unterberg A, Maier-Hauff K, Dautermann C (1986) Role of mediator compounds in secondary brain damage – current evidence. In: Baethmann A *et al* (eds) Mechanisms of secondary brain damage. Plenum Press, New York, pp 139–150

99. Ward JD, Miller JD, Choi SC (1986) Failure of prophylactic barbiturate coma in the prevention of death due to uncontrollable intracranial hypertension in patients with severe head injury. In: Miller JD *et al* (eds) Intracranial pressure VI. Springer, Berlin Heidelberg New York, pp 766–768

100. Yen MH, Whright D, Nakagawa H (1985) Effects of dexamethasone on the blood- brain distribution of (125-J) albumine and (14-C) alpha-aminobutyric acid in vasogenic cerebral edema. In: InabaY *et al* (eds) Brain edema. Springer, Berlin Heidelberg New York, pp 638–645

101. Yoshita S, Inoh S, Asano T (1986) Effect of transient ischemia on free fatty acids and phospholipids in the gerbil brain. J Neurosurg 53: 323–331

102. Yoshino E, Yamaki T, Higuchi T (1985) Acute brain edema in fatal head injury: analysis by dynamic CT scanning. J Neurosurg 63: 830–839

103. Zimmerman RA, Bilaniuk LT, Bruce D (1978) Computed tomography of pediatric head trauma: Acute general cerebral swelling. Radiology 126: 403–408

104. Zimmerman RS, Muizelaar JP, Wei EP, Kontos HA (1989) Reduction of intracranial hypertension with free radical scavengers. In: Hoff JT, Betz AL (eds) Intracranial pressure VII. Springer, Berlin Heidelberg New York, pp 804–809

105. Zülch KJ (1967) Neuropathological aspects and histological criteria of brain edema and brain swelling. In: Klatzo I *et al* (eds) Brain edema. Springer, New York

106. Zülch KJ (1986) Brain tumors. Their biology and pathology, 3rd ed. Springer, Berlin Heidelberg New York

Posttraumatic Intracerebral Hematomas

G. FOROGLOU

Department of Neurological Surgery, "Ahepa" General Hospital,
University of Thessaloniki (Greece)

With 12 Figures

Contents

"Time has not yet come – and it is questionable if it will ever come – when
the diagnosis concerning the localization of a haemorrhage can be made

with such a precision that the success of the trepanation, in case of apoplexia, will be totally secured"

Starr: Hirnchirurgie, Leipzig, 1894

We should be thankful that this historic statement is no longer valid.

Definition

A cerebral hemorrhage must be distinguished from a hematoma on several grounds.

Anatomo-pathologically a hemorrhage is a more or less diffuse extravasation infiltrating the cerebral tissue with consequent destruction of it; a hematoma is a well circumscribed area of bleeding, mostly expanding into the white substance, which splits by pushing fibre tracts aside and causing compression of the surrounding structure (Bagley 1932).

Concerning the topography it seems that the hemorrhage has a deeper location, while the hematoma has a more superficial one, into a cerebral lobe, sometimes reaching the cortical surface (Samiy 1962).

As Luyendijk (1972) pointed out, the term hematoma is not admitted by all authors writing on this subject; instead of hematoma some authors use the term "massive hemorrhage".

After the proposition of Russel (1954) it is accepted world-wide that an intracerebral bleeding is considered to be a hematoma or massive hemorrhage when its size exceeds 3 cm in diameter: if it is localized into the brain stem its critical size is 1.5 cm in diameter.

Aspects of their diagnosis and treatment are unique, and they have important medicolegal connotations, especially the so-called "traumatische Spät-Apoplexie" or "delayed traumatic hematoma".

Traumatic intracerebral hematomas are rarely solitary lesions; they are usually accompanied either by another type of intracranial hemorrhage (epi- or subdural) or by a contusion and cerebral edema (Cassassa 1924).

The associated lesions may lie on the same side, contralaterally, or in the brain stem, and may occur in many different combinations and variations. In the acute phase it is very difficult to diagnose clinically and to separate from contusions. The latter are commonly associated with local hematomas, so in practice their division is arbitrary. All these complexities and problems have made it difficult to collect reliable statistics about their occurrence before the availability of CT scan (Luyendijk 1972).

Incidence

It is generally accepted that the incidence of posttraumatic intracerebral hematomas is very low but highly variable, and depends on the overall

Table 1. *Author's Series (over a two-year period) of 276 Traumatic Intracerebral Hematomas*: Sex Distribution, Males: Females = 3:1

Male	210	76%
Female	66	24%
Total	276	100%

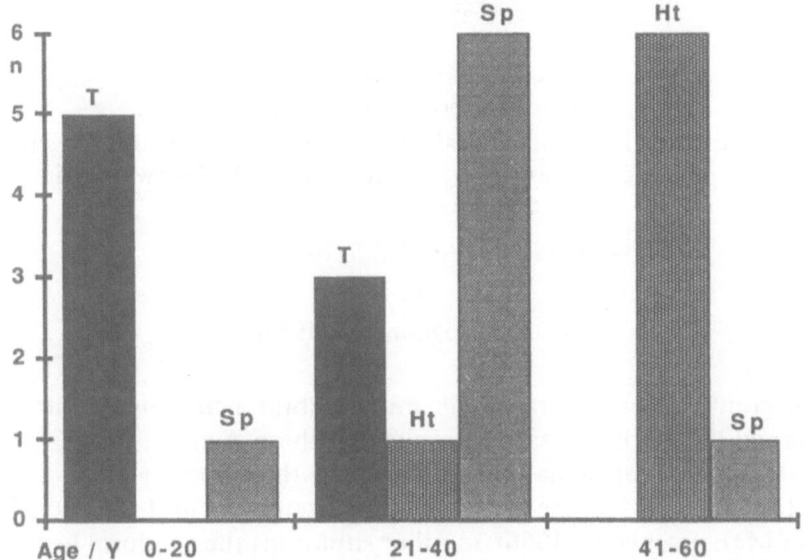

Fig. 1. Correlation between the age of the patients and the kind of hematomas
T Trauma; *Sp* Spontaneous; *Ht* Hypertension

material of each author. It varies from 0.35% in Lin *et al.* (1958), 1.3% in Dietz (1962) and Afra *et al.* (1963), 1.6% in Gurdjian *et al.* (1958), 2% in Loew *et al.* (1960), Busch (1963), Dacey *et al.* (1986), 2.5% in Bochiardo *et al.* (1972), 3% in Alayza Escarod *et al.* (1956), 5% in Frowein (1965), 5.4% in Krauel (1966) up to 16% in Luyendijk (1972). In our own material, consisting of 276 traumatic intracranial hematomas of different kind, the incidence of the 17 intracerebral hematomas was 0.2% (Foroglou *et al.* 1986). The correlation males to females was 3 to 1 (Table 1).

Age distribution of traumatic and nontraumatic hematomas is represented in Figs. 1 and 2.

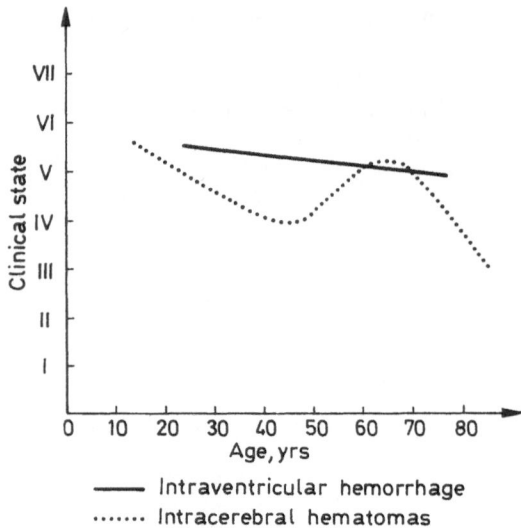

Fig. 2. Age of the patients and clinical state in intracerebral and intraventricular hematomas. The clinical states I to VII correspond to the following points in the Glasgow Coma Scale:

I   = 13–15;   II = 10–12;   III = 9–10
IV = 7–8;   V = 6–7;   VI = 4–5;   VII = 3

(after Foroglou *et al.* 1986)

The combination of traumatic intracerebral hematomas with other intracranial lesions has greatly complicated these studies (Table 2).

Often a subdural hematoma is found in the direct vicinity of an intracerebral one: 62% of the cases stated by Stender *et al.* 1965, 50% in the series of McLaurin *et al.* (1956). In other situations the subdural hematoma may be found over the controlateral hemisphere. The most common combination of subdural and intracerebral hematomas has been found in temporal

Table 2. *Author's Series*: *Type of Traumatic Cranial/Cerebral Lesions*

Epidural hematomas	71	25.7%	1	delayed
Acute subdural hematomas	31	11.2%		
Chronic subdural hematomas	46	16.7%		
Intracerebral hematomas	17	6.2%	3	delayed
Intraventricular hematomas	4	1.4%		
Depressed fractures	38	13.8%		
Others	69	25.0%		
Total	276		4 = 1.4%	

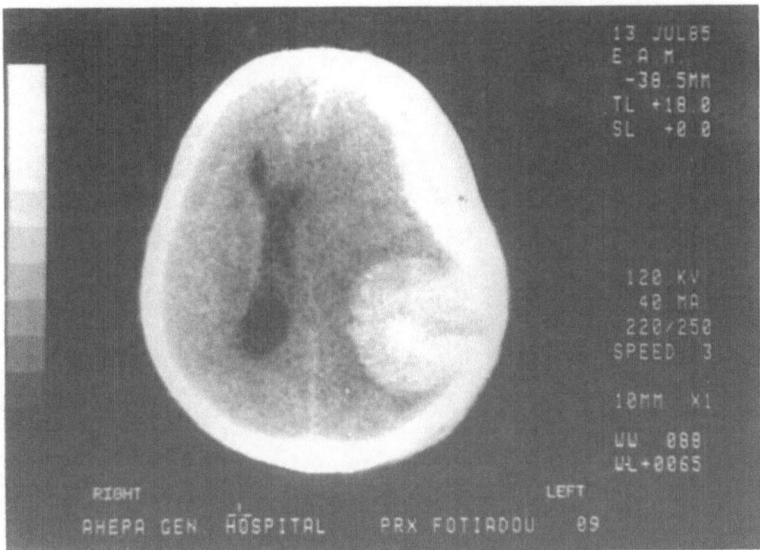

Fig. 3. Combined subdural and intracerebral hematomas

lobe lesions (Fig. 3). With an epidural hematoma the frequency is less and was found in only 2.5% of the cases (Fig. 4). Stender *et al.* (1965), Tönnis *et al.* (1963) and Samiy (1962) reported that all three types of hematoma may coexist in the same patient.

The mortality may exceed the 58% reported by Karimi-Nejad *et al.* (1979), depending on the severity of the group studied. In our own material this percentage was even higher up to 71%.

Kretschmer (1979) remarks that more than 30% of intracerebral hematomas have associated subdural or epidural hematomas, without any relation to the intraparenchymal collection. Skull fractures are also present in over 50% of the cases, twice as frequent on the side of the intracerebral hematoma than on the opposite side.

Cerebral contusions of varying severity and location are frequently seen indeed in almost every case of traumatic intracerebral hematoma. They may occur at a distance often contralaterally (Fig. 4), and are of very poor prognosis, when situated deeply in the brain stem (Bush 1963).

Etiology and Pathogenic Mechanism

A severe head injury can damage cerebral tissue, thus a contusion may result. This type of head injury has been accepted by some authors to be "minor" (Barathan *et al.* 1972, Morin *et al.* 1970), "considerable" (Austerheim 1957), or "variable" (Doughty 1938, Naffziger *et al.* 1928).

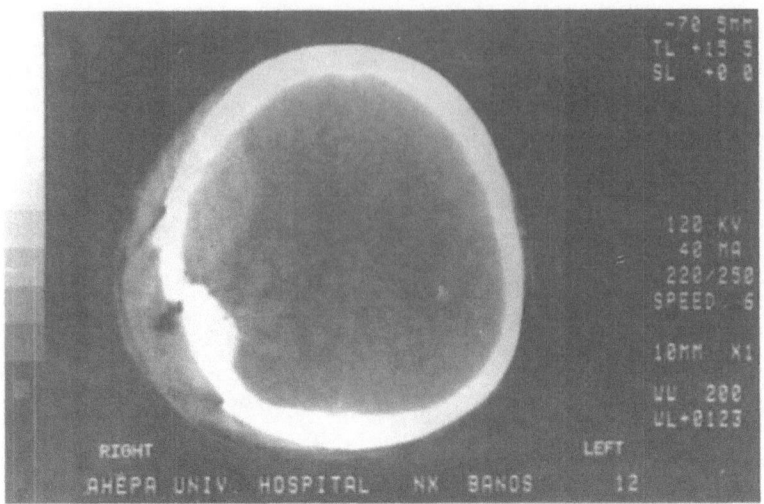

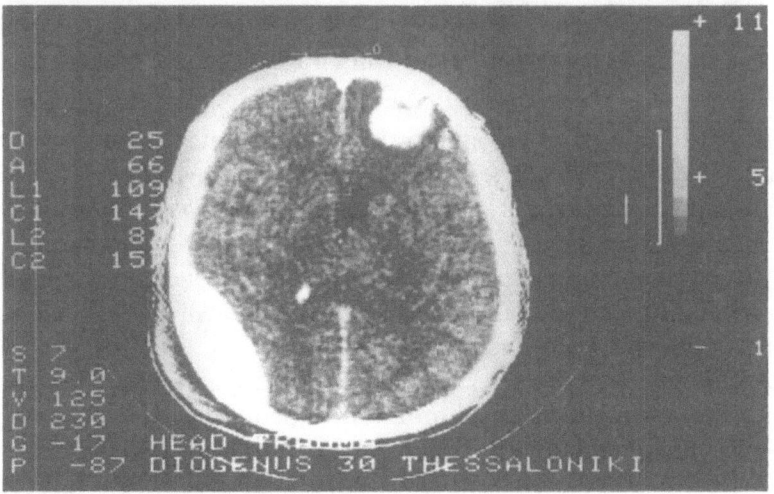

Fig. 4. Combined epidural and intracerebral hematomas. a) Depressed fracture associated with an epidural hematoma and next to them an intracerebral one. b) Epidural hematoma and "contre-coup" intracerebral hematoma

Mobility of the head at the time of impact has been considered to be an important factor in producing an intracerebral hematoma. The rotational forces that develop following an impact, when the head is not fixed, have been implicated in the development of parenchymal injuries of the brain by Courville *et al.* (1940), Gurdjian (1975), Levinthal *et al.* (1977), Koulouris *et al.* (1981).

As is well known, different factors would determine the degree of damage, depending on the violence, the rupture of small or even larger intracerebral vessels with, as a result, an intraparenchymatous hematoma.

The rapidity with which the hematoma becomes manifest, from a few hours to a few weeks, and its size will depend on many factors including the calibre of vessels and their types – artery or vein – this is of extreme importance.

The existence of a posttraumatic intracerebral hematoma in new cases and its correlation to trauma is easy to demonstrate, this relation must be obvious in cases appearing subsequently, because its proof may have an important medicolegal aspect (Luyendijk 1972, de Vet 1965, 1974).

To establish this relationship between cerebral trauma and intracranial hematoma, it becomes more and more difficult, as the interval between the trauma and the onset of its clinical effects lengthens in cases of "Spät-Apoplexie" or "Delayed intracerebral hematoma". On average the accepted interval has varied from a few days or weeks to several months. This point is disputed in the literature for many years by Marburg (1936), Naffziger et al. (1928), Symonds (1940), Antinnen et al. (1957), Zülch (1968, 1985, 1989), Nanassis and Frowein (1989).

Jewesbury (1947) points out that, in the absence of an accurate history, it may be almost impossible to differentiate traumatic "Spät-Apoplexie" from spontaneous intracerebral hematoma. Under such circumstances it may be impossible to decide whether trauma was the consequence of spontaneous "apoplectic intracerebral hematomas, or was the cause of an intracerebral hematoma".

Nelson et al. (1982) suggest the following pathogenesis to explain the progression of the contusion and the genesis of an intracerebral hematoma: "The initial event is sudden, and produces focal mechanical disruption of brain tissue, and vessels causing the initial petechial hemorrhages. Plasma leakage from these vessels and focal ischemia produce the circulatory alteration which leads to edema. The hemorrhage and edema, in turn, lead to a mass effect. Venous congestion which follows and dysautoregulation of regional blood flow lead to further edema and hemorrhage. Coalescent hemorrhages may produce further mechanical tissue disruption and a cycle of these pathophysiologic events may result in changes in blood pressure, oxygen- and carbon dioxide-pressure and local tissue metabolism and catecholamines are all additional factors that may influence the rate of progression" (Table 3).

Löfgren (1986) accepts that the primum movens of the bleeding is the transmural pressure in the vascular bed. He states "a main pathogenetic factor is probably loss of tamponading effect of a high ICP as a consequence of the removal of a mass lesion of intense hyperosmolar pressure or other pressure-reducing therapy. Coalescing hemorrhages may evolve in areas of brain contusion presumably related to local circulatory disturbances and vasoparalysis in a mechanically damaged vascular bed. These

Table 3. *Pathogenetic Schema for Delayed Effects in Cerebral Contusion* (after P. Nelson *et al.* 1982)

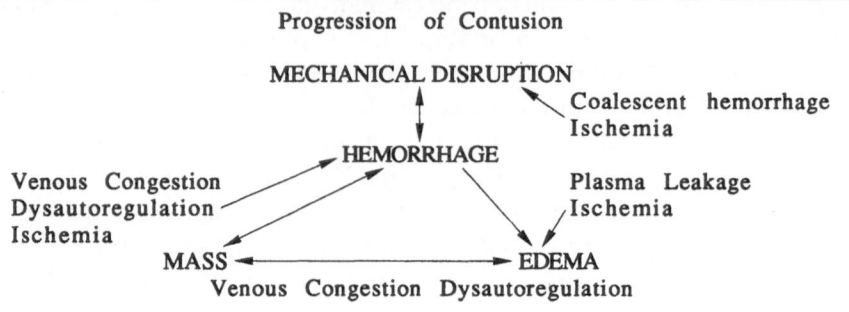

developments may be aggravated in the case of arterial hypertension acting on an unprotected vascular bed".

Other factors may be a disseminated intravascular coagulation-fibrinolysis syndrome (Brunetti *et al.* 1979; Pretorius *et al.* 1982).

Fukamachi *et al.* (1985) summarized the different etiologies of delayed traumatic intracerebral hemorrhages reported in the literature.

Biomechanics

It is known that a head injury is a complex ensemble of phenomena (Ommaya *et al.* 1976). A primary brain lesion, referring to visible structural disruption of neural tissue, correlates with the sites of concussion.

Although there are many detailed pathological studies, a comprehensive description of the relative three-dimensional distribution of all primary lesions and reactive changes, throughout the brain at known intervals of time, after head injury, is not yet available either in man or in animals.

Ripperger (1976), Sellier *et al.* (1965), Unterharnscheidt *et al.* (1965, 1966), Unterharnscheidt (1975), Sano (1965, 1972) and many others tried to explain with experimental studies and models that translation (linear) acceleration produces cavitation whereas rotational acceleration provokes coup and contre-coup lesions. For Unterharnscheidt *et al.* (1965) hemorrhages and necroses may be the consequences of cavitation.

In the earlier stage of the process continuity of tissue is fully preserved and lacerations are rare. This observation is in contrast to the view, that rupture of vessels necessitates interruption of tissue continuity.

For further details see the special chapters by Stålhammar and Nakamura in this volume.

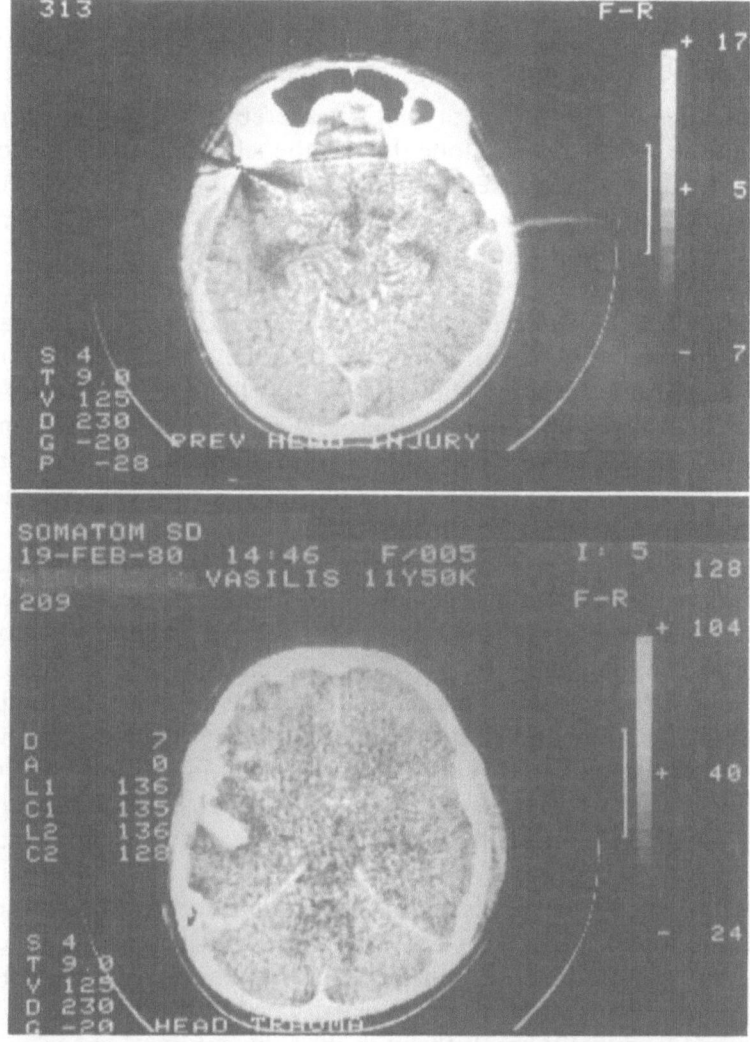

Fig. 5. Temporal intracerebral hematoma. Sometimes even a small intracerebral hematoma produces compressive signs and has to be treated surgically after a few days as in the case illustrated above. This example emphasizes that the size of a hematoma is not the only factor to be considered in the evolution and the prognosis. Every case has to be considered on its own

Location

Most authors agree that posttraumatic intracerebral hematomas are most frequently located in the temporal area (Fig. 5); frontal, parietal, occipital and cerebellar locations being exceptional, as reported by

McLaurin *et al.* (1956, 1965), Loew *et al.* (1960), Tönnis *et al.* (1963), Browder *et al.* (1942), Friedmann *et al.* (1960), Samiy (1962), Krauel (1966), Gurdjian (1975), Hamel *et al.* (1978), Kretschmer (1979).

Jamieson *et al.* (1972) accept that intracerebral hematomas are more frequent in the right temporal lobe than in the left (46/26). On the contrary Lin *et al.* (1958) reported an equal distribution of hematomas between the temporal and frontal lobes. Kretschmer (1979) and Demiere *et al.* (1987) found more hematomas on the left than on the right side.

Jamieson *et al.* (1972) believe that intracerebral hematomas tended to result from lateral blows rather than by axial ones. In 65% of the cases the blow was located to the contre-coup side. They specify that temporal hematomas are due to lateral blows in 61% of cases, with an ipsilateral blow in 58% of them; a frontooccipital blow caused a temporal hematoma in 20% of patients. The rare frontal hematomas are caused by an occipital blow in the majority of cases, 46%; only in 8% of them was the blow frontal, with an ipsilateral location in 64%.

Brzezinski *et al.* (1964), McPherson *et al.* (1986) and Courville *et al.* (1940) observed that hematomas very seldom occur in the internal capsule, and Kaufmann (1929) states that "it is rare to find a traumatic hemorrhage in the basal ganglia, where the usual arteriosclerotic hemorrhages are so frequent". MacPherson *et al.* in a recent article (1986) review the subject of intracerebral location and found that 2% of cases had a basal ganglia hematoma. These patients were more severely injured and they usually share many features with those patients with a diffuse white matter lesion and have a worse prognosis than other traumatic intracranial hematomas (Fig. 6).

Kretschmer (1979) remarks that deep, primary central hematomas are the result of direct trauma with tearing of veins or arteries. The subcortical intracerebral hematomas are considered to be due to cortical contusion progressing to a hematoma. They are usually no larger than 3 centimeter in diameter and frequently break through into the subdural space.

Rapoport *et al.* (1980) found, to their mind, a most unusual location of hematoma in the corpus callosum.

Other locations concern intraventricular hematomas (Fig. 7a,b). Zuccarello *et al.* (1981) point out that less than 3% of all patients with blunt head injuries had intraventricular hematomas. For Cordobes *et al.* (1983) this percentage varies from 1.5–5.7%. The relationship to intraparenchymal hematomas is 1/3.

The pathogenesis of these intraventricular hematomas is not yet clear, the following causes are discussed:

When the intraventricular hematoma follows or is associated with an intracerebral hematoma, it can be postulated that there is a dissection of

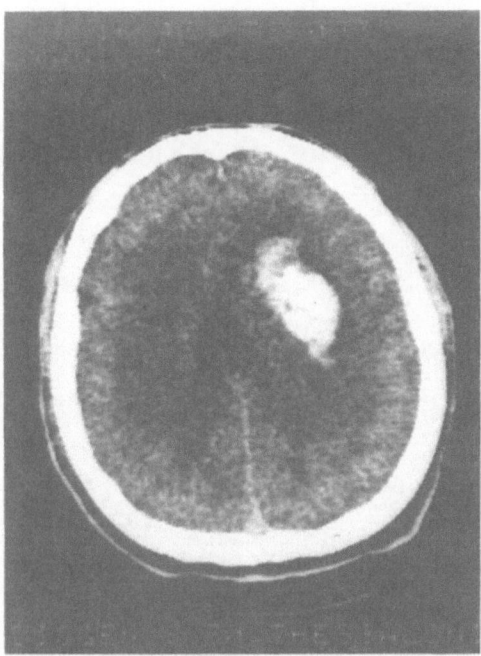

Fig. 6. Traumatic intracerebral hematoma deeply situated in the basal ganglia on the right side. Case of a male 20 years old victim of a traffic accident three hours before admission. On clinical examination superficial bruises of the scalp in both temporal regions, deep coma, mydriasis both side, areflexia, disturbances of respiration, exitus next day

blood from an intracerebral hematoma through the ependymal lining with extension of blood into the ventricular system. It must be remembered that brain injuries by themselves can lead to severe disturbances of blood coagulation in previously normal patients. The effect of brain trauma on blood coagulation may be local or there can be general effects such as disseminated intravascular coagulation (Pretorius and Kaufmann 1982).

Another cause may be the rupture of an unsuspected arteriovenous malformation (Hodge et al. 1975).

When the intraventricular hematoma is solitary or when there is no extension of an intracerebral hematoma into the ventricular system, another mechanism could be postulated: the primary lesion in the midline (3rd ventricle, corpus callosum) results from a development of negative pressure in this region. The ventricular walls and the portion of the corpus callosum adjacent to the ventricles are particularly subject to the effects of the negative pressure and are distorted: When a blow is applied along the sagittal diameter of the skull the resulting deformation will consist of an increase in the minor axis and a decrease of the major axis of the ellipsoidal

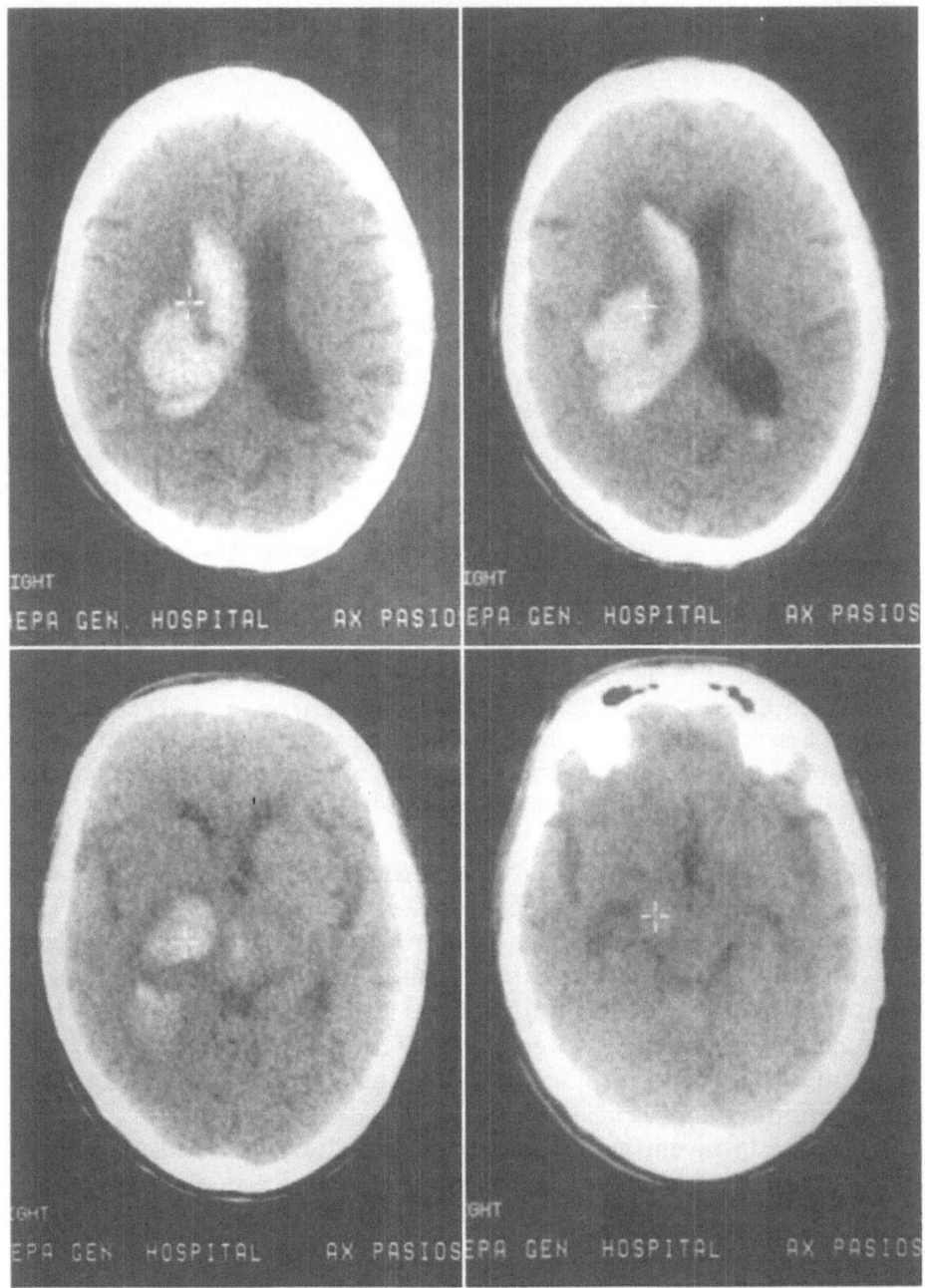

Fig. 7a

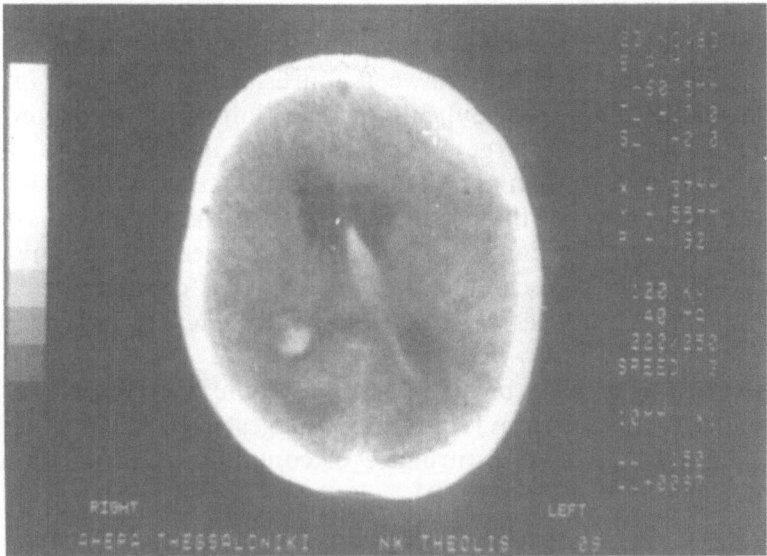

Fig. 7a and b. An intraparenchymal hematoma is rarely associated with an intraventricular one (a); even more rare is the presence of an epidural hematoma found
with intraventricular ones (b)

skull. These changes will result in an increase in total volume with a dilatation of the ventricles. The ventricular walls contain the so-called subependymal veins that lie in all parts of the ventricular system. Rupture of the
ventricular walls could be the origin of the solitary ventricular hemorrhage
(Unterharnscheidt *et al.* 1966, Lanksch *et al.* 1976).

Prior to the advent of CT, the diagnosis of an intraventricular hematoma could be made with good probability in some cases, with involvement
of the 3rd and 4th ventricles (Pia 1968, 1980): These cases show a classical
clinical picture described by Sanders (1881) (cited by Pia), consisting of deep
coma of acute onset, focal deficit symptoms of mesencephalic herniation
and central regulation disorders; death may occur in 24 hours as a result of
respiratory paralysis. Janny *et al.* (1966), Cordobes *et al.* (1983), Heiden
et al. (1983), Ruscalleda *et al.* (1986) in their extensive studies accept that
intraventricular hematomas are in more than 50% of cases associated with
diffuse brain lesions, intracerebral contusions and other extra- or intra-
axial hematomas. All patients were in deep coma and very few survived.

Clinical Manifestations

The majority of patients with intracerebral hematomas have suffered
a more severe cerebral trauma than do patients with extracerebral collections

(French *et al.* 1977). The consequence being that they are deeply comatose and have a low score in the Glasgow coma scale.

The clinical differentiation between contusional lesions and intracerebral hemorrhage can be very difficult and even impossible to diagnose only on a clinical basis. Even large intracerebral hematomas can be clinically completely silent. Barathan *et al.* (1972) reported that the larger intracerebral hematoma appeared to be associated with longer asymptomatic intervals. Huttarsch and Cardanus (1978) report on cases without any clinical manifestation. For Kretschmer (1979) there is no characteristic clinical picture and for Yashon *et al.* (1978) the neurological symptoms may be acute, intermittent and/or progressive.

As there is no clinical picture pathognomonic of intracerebral hematoma, some circumstances and signs might suggest it. In acute cases, it is the occurrence of progressive disturbance of consciousness with or without localizing signs.

Frowein (1965) in their report found that a lucid interval was present in 40% of patients, atypical symptoms in 14% and typical ones in 15% of the material studied and collected from different centers in Europe.

Localizing Signs

I. Cranial nerves

Unilateral pupillary dilatation either from the onset, or found subsequently is common. In severe cases both pupils may be dilated at the time of the first examination. Recently Katayama *et al.* (1985) reported on a case of traumatic homonymous hemianopsia associated with a juxtasellar hematoma, and Levin *et al.* (1985) reported on anosmia, as the only signs of an intracerebral hematoma.

II. Long tracts signs

These signs may be present from the beginning, but they can be delayed.

III. Papilledema

Reported by Lazorthes (1956), Gurdjian *et al.* (1958), Loew *et al.* (1960), Samiy (1962) and other authors in 5–18% of cases but never seen in the very acute ones.

IV. Progressive slowing of the pulse rate

Was described by Cushing in 1903 in his classic work.

Differential Diagnosis

It is well known that cerebral contusion, edema and diffuse hemorrhage may all give rise to the same symptoms and signs (Frowein 1965).

Traumatic carotid artery occlusions may also produce a similar clinical picture (Olafson and Christoferson 1970).

Previously angiography could only assist in the diagnosis, while today CT reveals the diagnosis precisely.

Occasionally craniocerebral trauma may cause a cerebral hematoma in an asymptomatic patient with a cerebral tumour (Goutelle *et al.* 1969). Hilton-Jones *et al.* (1985) looking after the different causes of stroke in the young, found "trauma to be the commonest identifiable predisposing factor for cerebral infarction in 22% and cerebral hematoma in 20%".

Intracerebral Hematomas Concerning Their Appearance in Time

Clinical and modern neuroimaging observations show that an intracerebral hematoma is rarely found on the first CT, performed a few minutes after a head injury. This will be called for practical purposes "early hematoma".

By far the majority of intracerebral hematomas produce clinical symptoms and can be visualized on repeat CT scans 8 to 48 hours after trauma: see the chapter by Frowein *et al.* in this volume.

Except for the early type there are hematomas appearing later on in the cerebral parenchyma. We will call them "late or delayed hematomas". These late hematomas were seldom diagnosed previously but with the improvement of our diagnostic tools they are found more frequently.

Their pathogenesis is still frequently discussed and the theories about their formation are controversial. Whatever their cause, the formation of a delayed intracerebral hematoma takes time, because of the different factors necessary for their development. It is not known when these factors supersede the mechanical forces of trauma.

One can distinguish two types of delayed intracerebral hematoma: one type appearing a few hours or days after the initial injury (Fig. 8) and a second type associated or appearing after a decompressive operation has been performed. The role of any surgical treatment of an extracerebral hematoma concerning damaged or contusioned brain is most important, and probably the significance of a delayed hematoma has to be considered as an evolution of the "immediate chronic type" of Jamieson, 1971, Piek and Bock 1986, Foroglou *et al.* 1986, and many other authors.

Fukamachi *et al.* (1979, 1985) describe 4 types of developmental processes of delayed intracerebral hematomas:

 I. 39% already present on initial CT
 II. 11% small or medium size initially and increasing afterwards
III. 24% present without changes on a second CT
IV. 26% initial CT salt and pepper appearance.

Young *et al.* (1984), on the basis of 15 cases underline the highly unpredictable development of intracerebral hematomas.

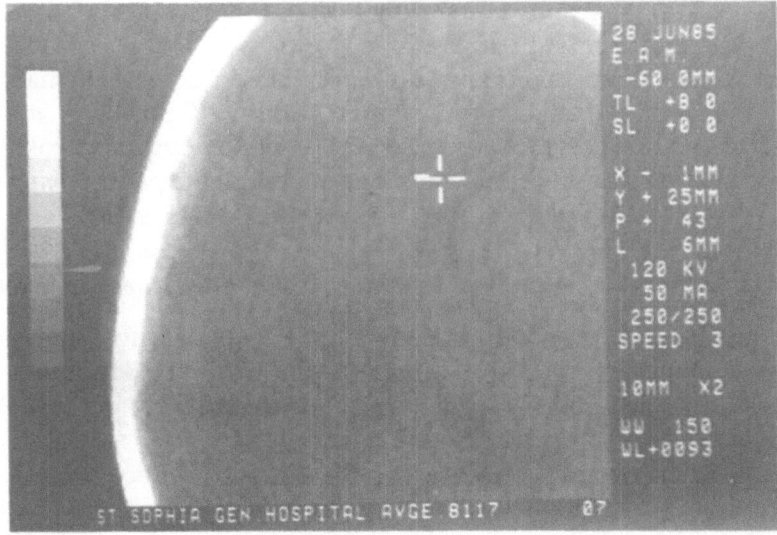

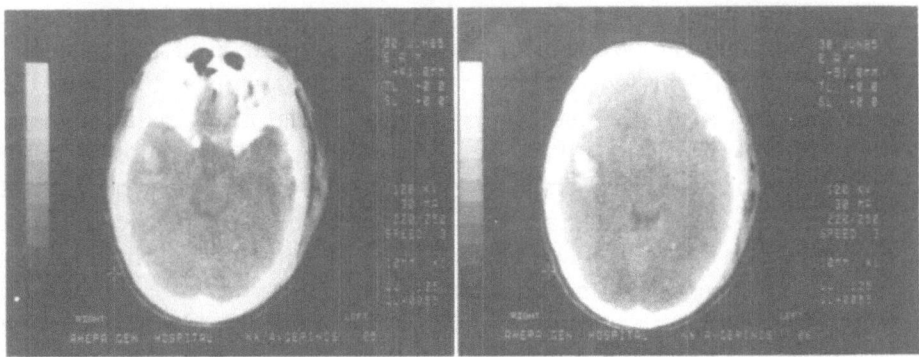

Fig. 8a–c. Delayed appearance of an intracerebral hematoma. On the first CT there are no parenchymal lesions seen except a profuse edema and a small epidural hematoma. A temporal intracerebral hematoma was found on the next CT two days later

Intracerebral Bleeding in Children

Because of the developing brain, the reactions to brain damage is more severe in children than in adults (Pia 1964). The degree of immaturity seems to parallel the tendency to necrosis of the damaged structures, but on the other hand this immaturity provides more possibilities for morphological and functional adaptation (Hallervorden 1939).

For Nakamura *et al.* (1981) the clinical outcome in newborn infants with hemorrhage depends on location, severity of bleeding, timing of the surgical treatment, associated brain damage, brain edema, hypoxia or respiratory distress.

The frequency of intracerebral hematomas in children is very low; Pia (1964) points out that it amounted to 1/10 of all hematomas studied. Hemorrhages resulting from birth vary considerably in the literature from 65% for any kind of cerebral lesion (Schwartz, 1924), to 2.1–2.7% of real parenchymatous lesions (Daamen 1956). Lassiera (1972) considers that the entity of intracerebral hematoma is very rare: 1.2% of all intracranial hematomas. Monges *et al.* (1972) share the same view, they found that 4% of intracranial hematomas in children have an intracerebral location, extradural 44%, subdural 24% and subdural–extradural 28%. Craft (1973) accepts that intracerebral hematomas are the least common type of hemorrhage requiring surgical treatment. This view is also shared by Wilberger *et al.* (1981), Choux (1984), Gjerris (1986).

Yoshida *et al.* (1979) reporting on hemophilic infants who had hematomas insist on their treatment with an antihemophilic agent.

Paraclinical Investigations

Radiography of the Skull

In acute and subacute cases a high incidence (90%) of linear fracture of the vault or the base of the skull has been reported. Stender *et al.* (1965) noticed that in 30–35% the fracture was on the controlateral side.

Hamel and Karimi-Nejad (1978) found fractures of the skull in 61% of their material, in 55% of the cases the fractures were homolateral to the hematoma and in 40% contralateral to it; finally in a few cases there was no fracture at all.

For Jennett (1980) head injured patients who develop serious secondary intracranial complications generally have fractures of the skull.

Demiere *et al.* (1987) accept that head injury cases with a skull fracture are at higher risk (p 0, 01) to develop intracerebral hematoma, and it is unusual to have an intracerebral hematoma with a depressed skull fracture.

Pneumoencephalography, Ventriculography

No longer used in head injury cases: These investigations have only a historical interest.

Cerebral Angiography

When CT or MRI are unavailable, cerebral angiography provides useful diagnostic information enabling neurosurgeons to proceed with therapy (Cooper *et al.* 1978).

Angiography is then of outstanding diagnostic value and must not be omitted, except for a few very urgent cases. It is of particular value to localize hematomas, but not for differential diagnosis, because other conditions such as temporal lobe contusions with edema, abscesses, cystic tumors or small hematomas in the frontal and temporal regions may give rise to a similar angiographic picture. That is an important notion because of the difference of treatment (Friedmann *et al.* 1960, Samiy 1962, Tönnis *et al.* 1963, Vigouroux *et al.* 1981).

Isotope Scintigraphy

Ojemann *et al.* (1965), Riccobono *et al.* (1970) reviewed its value. The method has little use in acute cases; in chronic cases it has been found to be of value in differential diagnosis. This method is no longer in use.

Echoencephalography

The favored method for head injury cases, a few years ago, has lost all of its interest after the availability of CT. The great advantage of this technique was its simplicity and its lack of risk. It took little time to carry out and could be repeated as often as necessary.

Many authors have stressed its diagnostic value: (De Vet 1965, 1974, Schiefer *et al.* 1967, Kanaya *et al.* 1968, Foroglou *et al.* 1970).

Like angiography it could not give any indication of the type of lesion, but was useful only for lateralisation, where it had been proved to be more reliable than the neurological signs. It has to be used in acute cases, before arteriography; it may be important not to lose time by performing arteriography on the wrong side. If the patient's condition deteriorates with a falling level of consciousness and increased shift of the midline echo on subsequent echoencephalograms operation must be performed at once.

Computerized Tomography

We all know from every day practice that CT has supplanted all previously cited methods in the diagnostic evaluation and management of head injuries. This relatively new investigation can demonstrate intracerebral hematomas even within one hour after the onset of symptoms (Pineda 1977), and differentiates them from other pathological entities; it can also give information about the size, the side, the extent and the surroundings of a hematoma. These traumatic intracerebral hemorrhages are shown in the CT scan as zones of increased density and are usually surrounded with a thin, less dense border. The density of such hematomas is between 25–35 EMI units, but may be greater because of hemoglobin aggregation.

Comparing the variations of the density of a hematoma with normal cerebral tissue the method can provide information about the progress of the resorption in the course of the disease (see the chapter of Frowein *et al.* in this volume).

After a time the abnormal areas become smaller, less dense and after 7 days or more the absorption coefficient is reduced to the level of CSF, resembling cystic cavities, or empty spaces filled with fluid such as porencephaly (Lennington *et al.* 1979).

The mass effect associated with intracerebral hematomas does not decrease as rapidly as the density of the lesion, which is more the result of degradation of hemoglobin and incorporation of fluid than actual resolution of the hematoma.

Multiplicity – Fig. 9 – has been also reported from 20–33%, (Moseley *et al.* 1976; Dublin *et al.* 1977; Zimmermann *et al.* 1977, 1980; Weisberg 1979). The size and the density usually decrease progressively with time (Dolinskas *et al.* 1977). However, enlargement of traumatic hematomas has also been observed (Davis *et al.* 1977).

The greatest diagnostic value of CT in cases of hemorrhage is in the acute case, where X-ray absorption is higher. With a low hemorrhage content the attenuation of extravasated blood may have the same value as brain tissue, and at very low hemoglobin concentration of the blood it may be less dense than brain tissue. This notion is of paramount importance in polytrauma cases where we have to deal with shocked and sometimes anemic patients.

Contrast enhancement occurs as a peripheral ring between the first and the sixth week post trauma in a pattern not always similar to that seen with infarction, abscess and neoplasm (Weisberg 1980). Zilkha (1983) reports on several cases associated with a fluid-blood level. This phenomenon has been

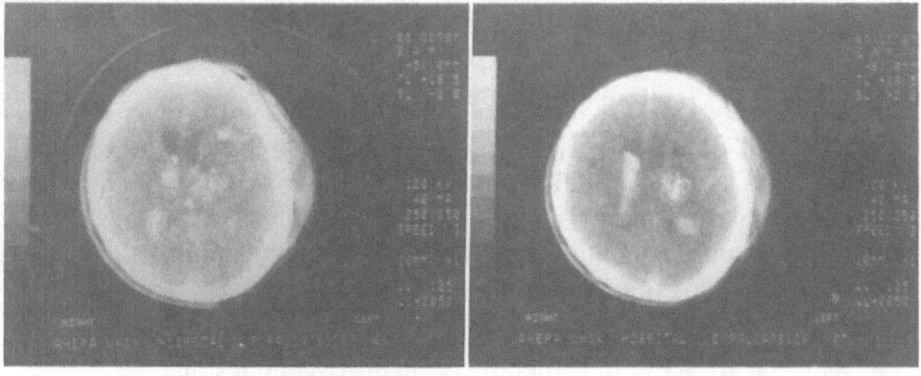

Fig. 9. Multiplicity of location of posttraumatic intracerebral hematomas

associated with hemorrhagic melanoma, or primary tumors of the CNS (Dublin *et al.* 1979; Zimmermann *et al.* 1980), or even in cases of recent AVM rupture (Richmond *et al.* 1981). Cystic cavity was not found at either surgery or autopsy and its mechanism is not known. But it indicates a quantity of freshly clotted blood and layers of unclotted blood or serum within the intracerebral hemorrhage.

Clinicopathological Correlation with CT

The clinico-morphological correlation made with the CT suggests that certain therapies should prevent deterioration of neurological function (Clifton *et al.* 1980). Early evacuation of focally damaged brain, as proposed by Becker *et al.* (1977), could decrease the incidence of late deterioration. We all know that compression of the cerebral tissue by hematoma or edema can produce ischemia. McPherson *et al.* (1986) have shown a high incidence of ischemic changes in the brain in fatally head injured patients associated with angiographic evidence of ischemia in life. There was a high correlation between intracerebral hematoma and local ischemic changes, supporting in a way the idea that mechanical distorsion, producing ischemia and infarction may cause delayed intracerebral hemorrhage.

Patients, who have hematomas or mass effect as seen on the initial CT scan, even if operated on, and patients who do not rapidly improve are at high risk of deterioration. From clinical observations there is no firm correlation between clinical picture, neurological deterioration, modifications of intracranial pressure, volume pressure response and CT findings (Robertson *et al.* 1979; Papo *et al.* 1979; Wozney *et al.* 1985). Papo *et al.* (1980) do not find any difference in the outcome of operated cases and disagree with Ugrumov *et al.* (1979) who postulate the contrary and operate on every case even those in deep coma, with good results. It is common sense that prompt operation in patients deteriorating from focal mass effect should decrease mortality, although the quality of life is an important and an unanswered question at present (Levin *et al.* 1978; Sweet *et al.* 1978).

Predicting Outcome with CT and MRI

With CT in the last 5 years or even more, it has been seen, in several countries, that there is a tendency to predict the outcome of severe head injury cases, based either on clinical data, or on CT scan findings. Lipper *et al.* (1985) proposed such a CT score (CTS) on posttraumatic hematoma cases, with the following equation:

$CTS = -0.62X1 + 1.32$, where
X1 is given by $X1 = 0$, if there is no hemorrhagic lesion,
$X1 = 1$ if this lesion is seen on one or two slices,

X1 = 2 if there are three or four slices,

X1 = 3 if there are seven or eight slices with hematoma material.

A patient with CTS of less than zero is predicted to have a poor outcome, otherwise the patient is predicted to do well. The size of hemorrhagic lesions on CT could also be used statistically to indicate prognosis: Stablein et al. (1980) and Narayan et al. (1981), used statistical models created for such purposes (Fig. 10).

Magnetic resonance imaging (MRI): Gandy et al. (1984) accept that magnetic resonance imaging provides better diagnostic informations than current CT techniques in some cases of head trauma.

Contusions may be better visualized on MRI images. This method can also demonstrate small isodense hematomas, that appear only as subtle abnormalities on CT, or isodense with the adjacent tissue, because of its superior contrast resolution. Hemorrhage is often a diagnostic dilemma in which clinically significant symptoms may be absent, equivocal, or even misleading. MRI may be relatively insensitive to hematoma in the first 24 hours after bleeding. Nonetheless considering these data more than 24 hours post trauma, MRI may have a definite advantage over CT in the detection of intracranial hematomas. This view is also shared by Swansen et al. (1985) and Gomori et al. (1985). Zimmermann et al. (1986) accept that acute hematomas are hyperdense and describe in detail 3 stages in the evolution of an intracerebral posttraumatic hematoma, as far as it concerns the center, the periphery and the adjacent cerebral tissue (see Table 4, Gomori et al. 1985).

Table 4. *MRI Score for Predicting Outcome* (after Gomori et al. 1985)

MRI: T_1-WI/T_2-WI *intensities compared with those of parenchyma*

Hematoma stage	Hematoma center	Hematoma Periphery	Rim of brain immediately adjacent to hematoma	White matter near hematoma
Acute	$=/\ll$ *	$=/=$	$=/=$	$=/>$
Subacute	$=/<$	$</=$ proceeding to $>/>$	$=/\ll$	$=/>$
Chronic	$>/>$	$>/>$	$=/\ll$	$=/=$

First symbol indicates T_1-WI intensity; second symbol indicates: intensity of T_2-WI

* Symbol: = indicates no change in intensity, > indicates increased intensity; < decreased intensity; ≪ markedly decreased intensity.

T_1-WI, T_2-WI: weighted images.

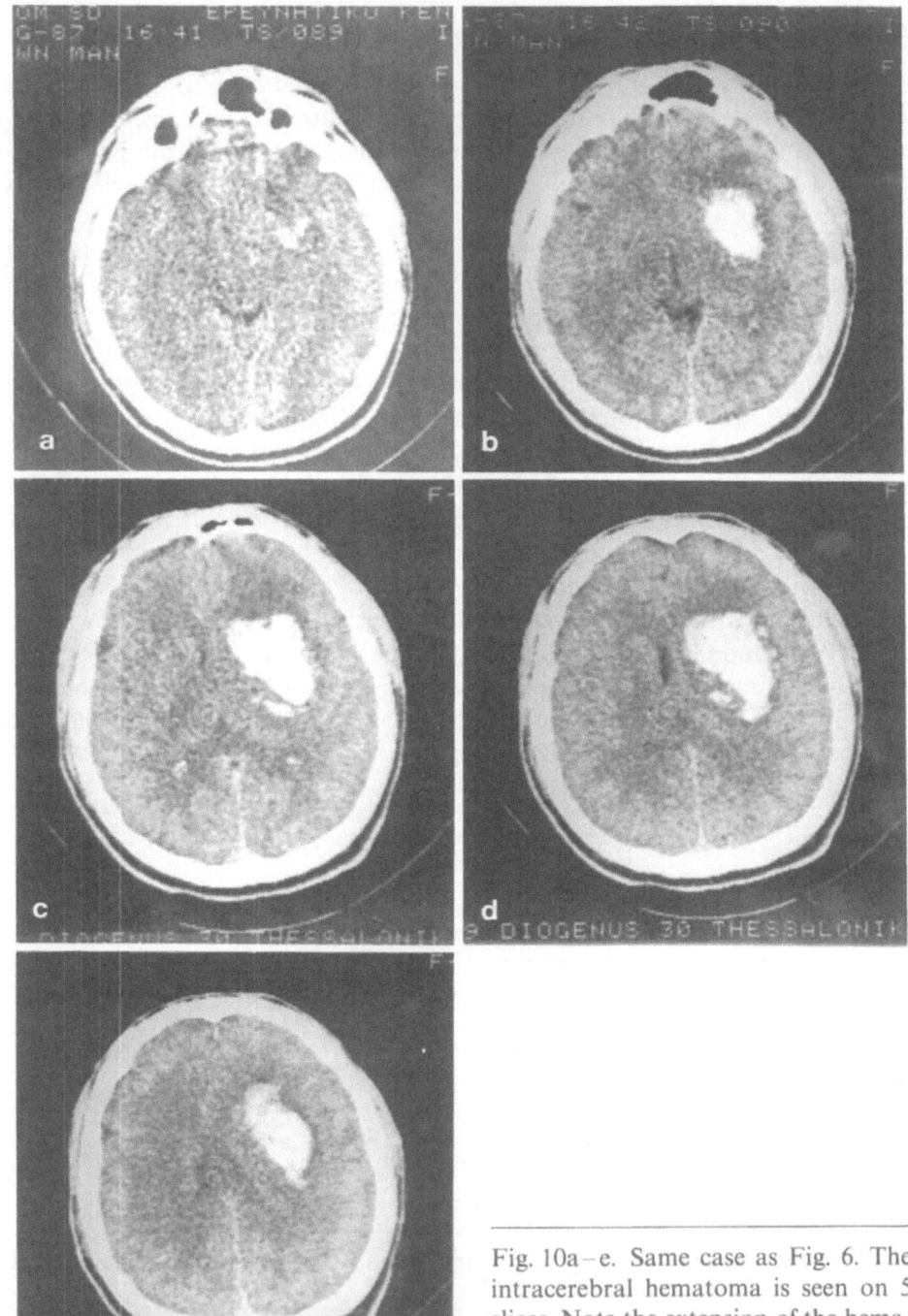

Fig. 10a–e. Same case as Fig. 6. The intracerebral hematoma is seen on 5 slices. Note the extension of the hematoma and severe cerebral oedema. Poor prognosis. Exitus next day

Treatment

The great majority of surgeons, if not all authors, believe that surgical treatment of intracerebral hematomas, acute or delayed, has a bad prognosis: Especially if the surgical intervention is performed very early and based principally on laboratory data rather than on clinical criteria (Fig. 10) (Browder *et al.* 1942, 1951, McLaurin *et al.* 1956, 1965, Ransohoff *et al.* 1965, Kalyanaraman *et al.* 1970, Levinthal *et al.* 1977, Kretschmer 1979, Heiden *et al.* 1983, Schester *et al.* 1985).

In the series of Jamieson (1971) and Jamieson *et al.* (1972) those patients who were conscious at the time of operation had a mortality rate of 6.2% compared with 45.2% for those who remained unconscious. Do the blood brain barrier and the cerebral blood flow play an essential role in such cases (Czernicki *et al.* 1977, Sussmann *et al.* 1974, Woodford *et al.* 1974) or are there any other factors playing a more prominent role in the success of such treatment? (Bose *et al.* 1982)

Whatever the factor is, it seems that there is no correlation between the size of the hematoma and the ICP level: there is probably CSF escape or increased absorption, or absorption of the injected blood.

The management of clinically silent posttraumatic intracerebral hematomas remains uncertain and controversial because every patient may deteriorate whether surgery is performed or not (Figs. 11 and 12). There are

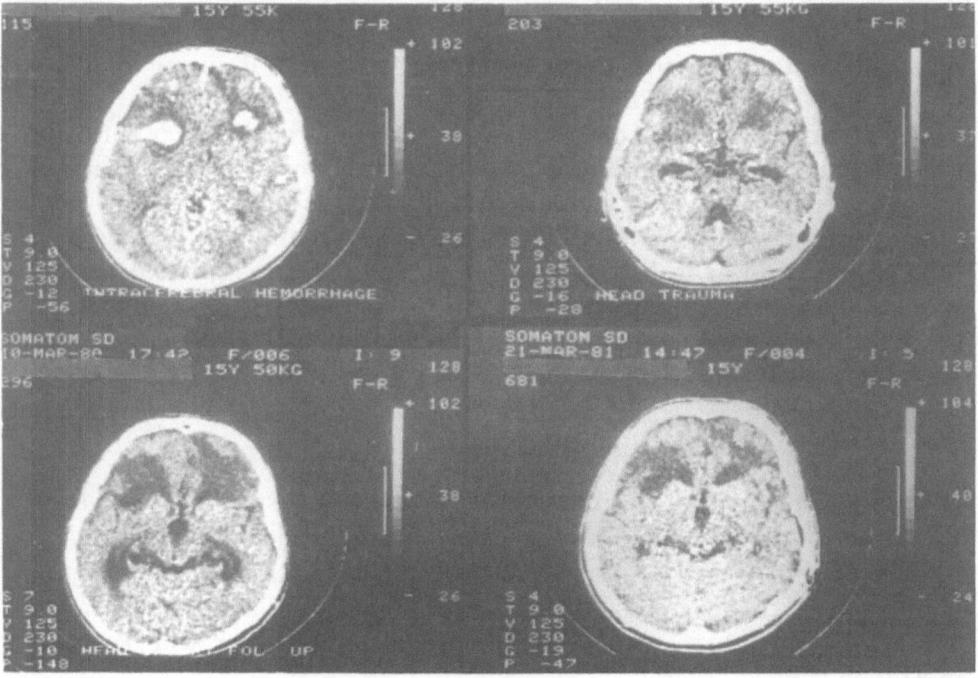

Fig. 11

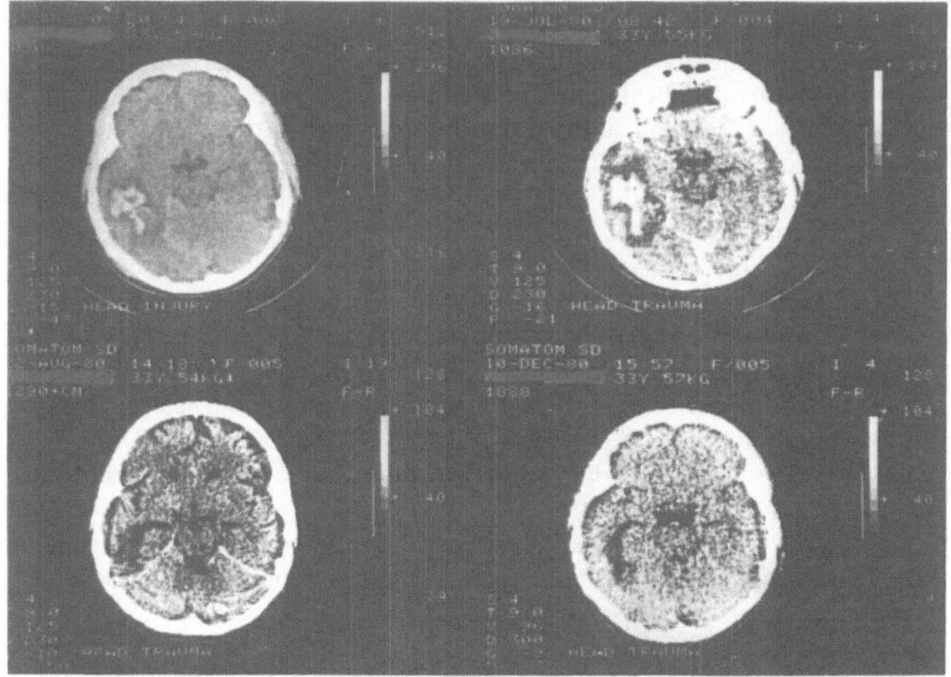

Fig. 12

Figs. 11 and 12. Evolution of posttraumatic intracerebral hematomas. Conservative therapy. Resorption without any sequelae

no reliable criteria for such treatment but the level of ICP during monitoring can provide some information (Teasdale *et al.* 1980).

Löfgren (1986) states that "while a delayed extracranial compressive lesion is usually an indication for operation, a conservative approach may be appropriate in the often indolent parenchymal lesions".

As far as delayed hematomas are concerned, craniotomy or craniectomy of burr-holes appear to be significant aggravating factors (p 0.05). Osmotherapy, barbiturate therapy or age do not appear to have influenced complications (Demiere *et al.* 1987). Young *et al.* (1984), however, advance a contrary theory on this last point.

How to Do It

The management of an intracerebral hematoma is still being discussed by various authors: Conservative therapy versus surgical intervention.

These controversial points of view have nothing to do with the "surgical temperament" of the neurosurgeon, nor the hematoma's appearance on the CT, nor of the values of the ICP measurements, but they will be based mostly on clinical criteria.

It is well understood that a head-injured patient has to be continuously monitored, his ICP measured, and be carefully observed in an Intensive Care Unit.

Many studies have shown that there is no absolute parallelism between the CT findings, the ICP values and the evolution of such posttraumatic cases.

It is accepted worldwide that the prognosis of an intracerebral hematoma is poor, whatever treatment is carried out. Our own material shows a mortality rate of 71%; we have to stress that the cases admitted to our department are selected from many city and country hospitals with a low Glasgow coma scale score of less than 7.

On admission of such a head injury patient we proceed, as usual, with the clinical assessment, the follow up of the vital functions, the neurological status and the X-ray investigations: at least skull, cervical spine and thorax, always under continuous medical observation. Afterwards the patient is taken to the CT department for a Scan, before admitting him into the Intensive Care Unit.

Following the results of the CT scan he is given antihypertensive drugs (like Mannitol, being aware that profuse quantities of that solution can produce many side effects), antiepileptic drugs, sedatives (if there is agitation) or antibiotics if necessary. Then he will be monitored by the medical and nursing staff for days, if necessary.

- If his clinical state shows a gradual amelioration he leaves this unit in a few days.
- If the patient's clinical state does not show any change at all, we proceed after 24, or more often after 48 hours, to carry out a further CT scan and reevaluate our treatment.
- If the clinical state shows deterioration inexplicable from the evaluation of his vital functions or other possible causes (in polytrauma cases), we proceed then to do a new CT scan, even earlier on, in order to rule out any secondary lesions.

According to the CT scan a decision is taken to continue the observations with conservative therapy, or not to do so.

If the patient needs surgical intervention, then he is brought to the theater at once.

If we have to deal with an intracerebral hematoma a craniotomy is performed over it, with an anterior decompressive craniectomy in cases with temporal location. We puncture the suspected area with a Cushing's needle, but sometimes the puncture is negative not because of misjudgement of location, but because the hematoma is not liquid. For that reason we then prefer to proceed immediately to make a small corticotomy, aspirate the clotted blood gently and look into the cavity for better hemostasis, always being careful to leave a thin wall of clotted blood over the cerebral parenchyma for hemostasis. Then we cover the cavity with a foil of hemostatic material (be aware not to put much of this material into the cavity, because of its high absorbing quality it can act locally as an expanding process).

If cerebral edema is important then we leave the dura open, or make a plastic operation with a substitute of dura. We insert a tube with continuous aspiration (Redon), and very often do not replace the bone flap to allow a better decompression. Then the patient is returned to the Intensive Care Unit.

References

1. Afra D, Vidovszky T (1963) Über traumatische intrazerebrale Hämatome. Zbl Neurochir 24: 88–94
2. Alayza Escarod F, Polo P, Cesar DM, Vallenas (1956) Hematomas intracerebrales traumaticos. Rev Med Hosp Obrero (Lima) 5: 76
3. Antinnen EE, Hillblom E (1957) On the apoplectic conditions occurring as delayed symptoms after brain injuries. Acta Neurol Scand 32: 103
4. Austerheim K (1957) Delayed traumatic intracerebral hemorrhage (Bollinger's Spät-Apoplexie). Report of one case with necropsy. Acta Pathol Microbiol Scand 38: 177–185
5. Bagley C jr (1932) Spontaneous cerebral hemorrhage. Discussion of four types with surgical consideration. Arch Neurol Psych (Chic) 27: 1133–1174
6. Barathan G, Dennyson WG (1972) Delayed traumatic intracerebral hemorrhage J Neurol Neurosurg Psychiat 35: 698–706
7. Becker DP, Miller JD, Ward JD, Greenberg RP, Young HF, Sakalas R (1977) The outcome from severe head injury with early diagnosis and intensive management. J Neurosurg 47: 49–52
8. Bochiardo E, Beguelin S, Romero M, Vera A (1972) Hematomas traumaticos intracraneanos. Rivision de 256 casos. In: Carrea R, Christensen JC, Turjanski L (eds) Neurotraumatologia Proceedings of the International Conference on Neurotraumatology. Buenos Aires, August 27–29, 1972, pp 307–319
9. Bose B, Kraut W, Osterholm J (1982) Intracerebral hematoma: Spontaneous cure by drainage into the middle ear. Neurosurgery 10: 103–104
10. Browder EJ, Corradini EW (1951) Surgical treatment of intracerebral hematomas. AMA Arch Neurol Psychiat 65: 112–117
11. Browder EJ, Turney F (1942) Intracerebral hemorrhage of traumatic origin: its surgical treatment. NY State J Med 42: 2230–2235
12. Brunetti J, Zingesser L, Dunn J, Rovit RL (1979) Delayed intracerebral hemorrhage as demonstrated by CT scanning. Neuroradiol 18: 43–46
13. Brzezinsky J, Jagodzinski Z, Szapiro J (1964) Hematomes multiples intracraniens posttraumatiques. Neurochirurgie 10: 333–338

14. Bush EHY (1963) Brain stem contusions. Clin Neurosurg 9: 18–33
15. Cassassa CSB (1924) Multiple traumatic cerebral hemorrhages. Proc Path Soc NY, MS 24: 101–106
16. Choux M (1984) Head trauma in children. EANS Course in Neurosurgery, Edinburgh, personal communication
17. Clifton GL, Grossman RG, Makela ME, Niner ME, Handel S, Sadhu V (1980) Neurosurgical course and correlated computerized tomography findings after severe closed head injury. J Neurosurg 52: 611–624
18. Cooper PR, Moody S (1978) Neurodiagnostic studies and the management of head injuries Comput Tomograph 2: 197–206
19. Cordobes F, de la Fuente M, Lobato RD, Roger R, Perez C, Millan JM, Barcena A, Lamas E (1983) Intraventricular hemorrhage in severe head injury J Neurosurg 58: 217–222
20. Courville CB, Blomquist OA (1940) Traumatic intracerebral hemorrhage with particular reference to its pathogenesis and its relation to "delayed traumatic apoplexy" Arch Surg 41: 1–28
21. Craft AW (1973) Head injury in children. In: Vinken PJ, Bruyn GW (eds) Handbook of Clinical Neurology, Vol 23. North Holland Publ Co, Amsterdam-Oxford; American Elsevier Publ Co Inc, New York, pp 445–458
22. Czernicki Z, Kozniewska E (1977) Disturbances in the blood-brain barrier and cerebral flow after rapid brain decompression in the cat. Acta Neurochir (Wien) 36: 181–187
23. Cushing H (1903) The blood-pressure reaction of acute cerebral compression, illustrated by cases of intracranial hemorrhage. Amer J Med Sci 125: 1017–1044
24. Daamen CBF (1956) Draematurites en subpendymale en intraventriculaire bloedingen. Ned T Geneesk 100: 1205
25. Dacey RG, Alves WM, Rimel RW, Winn HR, Jane JA (1986) Neurosurgical complications after apparently minor head injury. J Neurosurg 65: 203–210
26. Davis KR, Taveras JM, Robertson GH et al (1977) Computed tomography in head trauma. Semin Roentgenol 12: 53
27. Demiere B, Schoenle PW, Hori A, Brunke J, Spoerri D (1987) L'hematome intracerebral secondaire post traumatique. Neurochirurgie 33: 12–16
28. De Vet AC (1965) Middle fossa intra- and extra-cerebral lesions. Their treatment and prognosis. In: Proceedings IIIrd International Congr Neurological Surgery, Copenhagen. Intern Congr Series 110: 236–241 Excerpta Medica Foundation, Amsterdam
29. De Vet AC (1974) Traumatic intracerebral hematoma. In: Vinken PJ, Bruyn GW (eds) Handbook of clinical neurology Vol 24. North Holland Publ Co, Amsterdam Oxford; American Elsevier Publ Co Inc, New York, pp 351–368
30. Dietz H (1962) Zur Frage der Früherkennung frischer, traumatischer intrakranieller Hämatome. Ber Unfallchir Mainz (1962): 137–153
31. Dolinskas CA, Bilaniuk LT, Zimmerman RA et al (1977) Computed tomography of intracerebral hematoma. II Radionuclide and Transmission CT studies of the perihematoma region. Am J Roentgenol 129: 689

32. Doughty RG (1938) Posttraumatic delayed intracerebral hemorrhage. South Med J 31: 254–256

33. Dublin AB, French BN, Rennick JM (1977) Computed tomography in head trauma. Radiol 122: 365

34. Dublin AB, Norman D (1979) Fluid-fluid level in cystic cerebral metastatic melanomas. J Comput Assist Tomogr 3: 650–652

35. Foroglou G, Patsalas J, Kontopoulos B (1986) CT in head injuries. When and why. Proceedings 15th Hellenic Surgical Congr, Thessaloniki

36. Foroglou G, Zander E (1970) Estimation du diamètre des processus expansifs intracérébraux profonds par l'echo-encephalographie "A". Rev Med Suis Rom 90: 153–162

37. French BN, Dublin HB (1977) The value of computerized tomography in the management of 1000 consecutive head injuries. Surg Neurol 7: 171–183

38. Friedmann G, Schmidt-Wittkamp E, Walter W (1960) Serienangiographische Befunde bei traumatischen intracerebralen Hämatomen. Acta Neurochir (Wien) 8: 70–80

39. Frowein RA (1965) Intracerebral hematomas. In: Vigouroux PR, Frowein RA (eds) Survey of the organization of services for acute head injury in Europe. Proceedings IIIrd Intern Congr Neurological Surgery, Copenhagen. Intern Congr Series 110: 61–66 Excerpta Medica Foundation, Amsterdam

40. Fukamachi A, Kohno K, Wakao T, Tasaki T, Koizumi H, Nagaseki Y (1979) Traumatic intracerebral hematomas. A classification according to dynamic changes on sequential CTS. Neurol Med Chir (Tokyo) 19: 1039–1051

41. Fukamachi A, Nagaseki Y, Kohno K, Wakao T (1985) The incidence and development of delayed traumatic intracerebral haematomas. Acta Neurochir (Wien) 74: 35–39

42. Gandy SE, Snow RB, Zimmerman RD (1984) Cranial nuclear magnetic resonance imaging in head trauma. Ann Neurol 16: 254–257

43. Gjerris G (1986) Head injuries in children. Special features. Acta Neurochir (Wien) [Suppl] 36: 155–158

44. Gomori JM, Grossman RI, Goldberg HI, Zimmerman RA, Bilaniuk LT (1985) Intracranial hematoma: Imaging by High-Field MRI. Radiology 157: 87–93

45. Goutelle A, Lapras C, Trillet M, Rambaud G, Leger G (1969) Les hématomes intracérébraux révélateurs d'une tumeur cérébrale. Neurochirurgia 12: 218

46. Guerrero Isla AI, Ruiz A, Roda Frado JM, Alvarez MP, Ortega FV, Blazguez MG (1986) Hematomas intracerebrales tardios posttraumaticos. Riv Clin Espan 178: 112–115

47. Gurdjian ES (1975) Impact head injuries. Mechanistic clinical and preventive correlations. Ch C Thomas, Springfield Ill, p 370

48. Gurdjian ES, Webster JE (1958) Head injuries. Little Brown and Co, Boston, p 269

49. Hallervorden J (1939) Kreislaufstörungen in der Ätiologie des angeborenen Schwachsinns. Z Neurol 167: 526–546

50. Hamel E, Karimi-Nejad A (1978) Traumatic intracerebral hematomas. Advances in Neurosurgery 5: 56–61

51. Heiden JS, Small R, Coton W, Weiss M, Kurze Th (1983) Severe head injury. Am Phys Therapy Ass 63: 1946–1951
52. Hilton-Jones D, Warlow CP (1985) The causes of stroke in the young. J Neurol 232: 137–143
53. Hirsch LF, Spector HB, Boddanoff B (1981) Chronic encapsulated intracerebral hematoma. J Neurosurg 9: 169–172
54. Hodge CJ, King RB (1975) Arteriovenous malformation of the choroid plexus. Case Report. J Neurosurg 42: 457–461
55. Huttarsch H, Cardauns G (1978) Clinically non manifest hematomas. Advances in Neurosurgery 5: 68–70
56. Jamieson KG (1971) A first notebook of head injury. Butterworth, London, ed 2
57. Jamieson KG, Yelland JDH (1972) Traumatic intracerebral hematoma. Report of 63 surgically treated cases. J Neurosurg 37: 528–532
58. Janny P, Montrieuil B, Tournilac HM, Chabannes J, Tourde J (1966) Les hématomes intracérébraux rompus dans les ventricules. Neurochirurgie 12: 459–472
59. Jennett B (1980) Skull X-rays after recent head injury. Clin Radiol 31: 463–469
60. Jewesbury ECO (1947) Atypical intracerebral hemorrhage. Brain 70: 274
61. Kalyanaraman S et al (1970) Traumatic intracerebral hematoma. Neurol India 19: 30–33
62. Kanaya H, Yamasaki H, Saiki I, Furukawa K (1968) The use of echoencephalography to differentiate intracerebral hemorrhage and brain softenings. J Neurosurg 28: 539–543
63. Karimi-Nejad A, Hamel E, Frowein RA (1979) Verlauf der traumatischen intrazerebralen Hämatome. Nervenarzt 50: 432–435
64. Katayama Y, Yoshida K, Ogawa H, Tsubokawa T (1985) Traumatic homonymous hemianopsia associated with a juxtasellar hematoma after acute closed head injury. Surg Neurol 24: 289–292
65. Kaufmann E (1929) Pathology for students and practionners. Philad Med 3: 1890
66. Koulouris S, Rizzoli H (1981) Delayed traumatic intracerebral hematoma after compound depressed skull fracture: Case report. Neurosurgery 8: 223–225
67. Krauel V (1966) Intrazerebrale traumatische Hämatome. Schweiz Arch Neurol Neurochir Psychiat 98: 2
68. Kretschmer H (1979) Traumatic intracerebral hematomas – Analysis of 88 operative cases. Neurochirurgia 22: 35–41
69. Lanksch W, Meese W, Kazner E (1976) CT findings in closed head injuries with special reference to contusion. In: Lanksch W, Kazner E (eds) Cranial computerized tomography. Springer, Berlin Heidelberg New York, pp 318–328
70. Lassiera PA (1972) Traumatismos craneoencefalicos en la infancia. In: Carrea R, Christensen JC, Turjanski L (eds) Neurotraumatologia, Proceedings of the

Intern Conf on Neurotraumatology, Buenos Aires, Argentina, 27.-29.8.1972 pp 265–283

71. Lazorthes G (1956) L'hémorragie cérébrale vue par le neurochirurgien. Rapport présenté à la réunion de la Societé de Neuro-Chirurgie de Langue Française. Masson, Paris

72. Lennington BR, Laster DW, Moody DM, Bell MR (1979) Pre-enhancement ring density in resolving intracerebral hematomas. Comput Tomogr 3: 105–109

73. Levin HS, Grossman RG, Rose JE et al (1978) Long term neurophysiological outcome of closed head injury. J Neurol Neurosurg Psychiatry 41: 122–127

74. Levin HS, High WM, Eisenberg HM (1985) Impairment of olfactory recognition after closed head injury. Brain 108: 579–591

75. Levinthal R, Stern WE (1977) Traumatic intracerebral hematoma with stable neurological deficit. Surg Neurol 7: 269–277

76. Lin TH, Cook AW, Browder EJ (1958) Intracranial hemorrhage of traumatic origin. Med Clin N Amer 42: 603

77. Lipper MH, Kishore PRS, Enas GG, da Silva AAD, Choi SC, Becker DP (1985) Computer tomography in prediction of outcome in head injury. AJNR 6: 7–10

78. Loew F, Wüstner S (1960) Diagnose, Behandlung und Prognose des traumatischen Hämatoms des Schädelinneren. Acta Neurochir (Wien) [Suppl] 8

79. Löfgren J (1986) Traumatic intracranial hematomas: Pathophysiologic aspects on their course and treatment. Acta Neurochir (Wien) [Suppl] 36: 151–154

80. Luyendijk W (1972) Intracerebral hematoma. In: Vinken PJ, Bruyn GW (eds) Handbook of clinical neurology, Vol 11, Part I. Amsterdam North Holland Publ Co, Oxford, American Elsevier Publ Comp Co Inc, New York

81. MacPherson P, Teasdale E, Dhaker S et al (1986) The significance of traumatic hematoma in the region of the basal ganglia. J Neurol Neurosurg Psychiatry 49: 29–34

82. McLaurin RL, McBride BH (1956) Traumatic intracerebral hematoma; Review of 16 surgically treated cases. Ann Surg: 143–294

83. McLaurin RL, Helmer F (1965) The syndrome of temporal lobe contusion. J Neurosurg 23: 296–304

84. Marburg O (1936) Die traumatischen Erkrankungen des Gehirns und Rückenmarks. In: Bumke O, Foerster O (eds) Handbuch der Neurologie, Vol 11. Springer, Berlin

85. Monges J, Carrea R (1972) Observaziones estadisticas sobre traumatismos encefalocraneanos en la infancia. In: Carrea R, Christensen JC, Turjanski L (eds) Neurotraumatologia, Proceedings of the Intern Conf on Neurotraumatology, Buenos Aires, Argentina, 27.-29.8.1972

86. Morin MA, Pitts FW (1970) Delayed apoplexy following head injury ("Traumatische Spät-Apoplexie"). J Neurosurg 33: 542–547

87. Moseley TF, Zilkha E (1976) The role of computerized axial tomography (EMI Scaning) in the diagnosis and management of craniocerebral trauma. J Neuroradiol 3: 277

88. Naffziger H, Jones OW (1928) Late traumatic apoplexy, Report of 3 cases with operative recovery. California west. Med. 29: 361

89. Nakamura S, Yamada H, Tanaka Y, Kageyama N, Haroyanagi M, Zuzuki T, Gohda M (1981) Neonatal intracranial hemorrhage. Child's Brain 8: 397

90. Nanassis K, Frowein RA, Karimi A, Thun F (1989) Delayed posttraumatic intracerebral bleeding. Delayed posttraumatic apoplexy: "Spätapoplexie". Neurosurg Rev 12: [Suppl 1] 243–251

91. Narayan RK, Greenberg RP, Miller JD et al (1981) Improved confidence of outcome prediction in severe head injury. A comparative analysis of the clinical examination, multimodality evoked potentials, CT Scanning and intracranial pressure. J Neurosurg 54: 751–762

92. Nelson PB, Rosenbaum AE, Moossy J, Marron JC (1982) Delayed deterioration in the syndrome of temporal lobe contusion: Evaluation by computed tomography (CT). J Trauma 22: 39–42

93. Ojemann RG, Aronow S, Sweet WH (1965) Scanning with positron-emitting radio-isotopes. J Neurosurg 22: 489–498

94. Olafson RA, Christoferson LA (1970) The syndrome of carotid occlusion following minor craniocerebral trauma. J Neurosurg 33: 636–639

95. Ommaya AK, Gennarelli Th (1976) Experimental head injury. In: Vinken PJ, Bruyn GW (eds) Handbook of clinical neurology, Vol 23. North Holland Publ Co, Amsterdam Oxford; American Elsevier Publ Co. Inc, New York

96. Papo I, Caruselli G, Luongo A, Scarpelli M, Pesquini U (1980) Traumatic cerebral mass lesions. Correlations between clinical and computerized tomography data. Neurosurgery 7: 337–345

97. Papo J, Janny P, Caruselli G, Colnet G, Luongo A (1979) Intracranial pressure time course in primary intracerebral hemorrhage. Neurosurgery 4: 504–511

98. Pia HW (1964) Die traumatischen Hirnblutungen des Kindesalters. Acta Neurochir (Wien) 11: 583–600

99. Pia HW (1968) The diagnosis and treatment of intraventricular hemorrhages. Prog Brain Res 30: 463–470

100. Pia HW, Langmaid C, Zierski J (1980) Spontaneous intracerebral haematomas. Springer, Berlin Heidelberg New York

101. Piek J, Bock WJ (1986) Secondary intraventricular haemorrhage in blunt head trauma. Acta Neurochir (Wien) 83: 105–107

102. Pineda A (1977) Computed tomography in intracerebral hemorrhage. Surg Neurol 8: 55–58

103. Pretorius ME, Kaufman HH (1982) Rapid onset of delayed traumatic intracerebral hematoma with diffuse intravascular coagulation and fibrinolysis. Acta Neurochir (Wien) 65: 103–109

104. Ransohoff J, Randt CT (1965) Profiles of fatal and non-fatal closed head injury In: Proceedings IIIrd Intern Congr Neurological Surgery, Copenhagen. Intern Congr Series 110: 137–145. Excerpta Medica Foundation, Amsterdam

105. Rapaport AM, Falvey CF (1980) Posttraumatic headache caused by hematoma of the corpus callosum. Headache 20: 279–281

106. Riccobono XJ, Chase SP (1970) Fortuitous scan documentation of the development of an intracerebral hematoma. J Neurosurg 33: 79–81
107. Richmond T, Viroponagse C, Serwar M, Kier EL, Rothman S (1981) Intraparenchymal blood-fluid levels: new CT signs of arteriovenous malformation rupture. AJNR 2: 577–579
108. Ripperger EA (1976) The mechanics of brain injuries In: Vinken PJ, Bruyn GW (eds) Handbook of clinical neurology, Vol 23, Part I. North Holland Publ Amsterdam
109. Roberson FC, Kishore PRS, Miller JD, Lipper MAH, Becker DF (1979) The value of serial computerized tomography in the management of severe head injury. Surg Neurol 12: 161–167
110. Ruscalleda J, Peiro A (1986) Prognostic factors in intraparenchymatous hematoma with ventricular hemorrhage. Neuroradiology 28: 34–37
111. Russel DS (1954) The pathology of spontaneous intracranial hemorrhages. Proc Roy Soc Med 47: 689–693
112. Samiy E (1962) Das traumatische intrazerebrale Hämatom. Schweiz Med Wschr 92: 1565–1568
113. Sano K (1965) Survey of the organization of services for the treatment of acute head injury in Japan. In: Proceedings of the IIIrd Intern Congr Neurological Surgery, Copenhagen. Intern Congr Series 110: 39–45 Amsterdam, Excerpta Medica Foundation
114. Sano K (1972) Correlations between dynamics and pathology in head injury. In: Carrea R, Christensen JC, Turjanski L (eds) Proceedings of the Intern Conf on Neurotraumatology, Buenos Aires, Argentina, 27.-29.8.1972
115. Schester WP, Peper E, Tuatoo Y (1985) Can general surgery improve the outcome of the head injury victim in rural America? A review of the experience in America Samoa. Arch Surg 120: 1163–1166
116. Schiefer W, Kazner E (1967) Klinische Echo-Enzephalographie. Springer, Berlin Heidelberg New York
117. Schwartz PH (1924) Traumatic injury of the brain at birth. Dtsch med Wschr 50: 1375
118. Sellier K, Unterharnscheidt F (1965) The mechanics of the impact of violence on the skull. In: Proceedings IIIrd Intern Congr Neurological Surgery, Copenhagen. Intern Congr Series 110: 87–92. Excerpta Medica Foundation, Amsterdam
119. Stablein DM, Miller JD, Choi SC, Becker DP (1980) Statistical methods for determining prognosis in severe head injury. Neurosurgery 6: 243–248
120. Starr A (1984) Hirnchirurgie. Deuticke, Leipzig
121. Stender A, Schulze A (1965) The surgical treatment of space-occupying contusions and intracerebral hematomas after blunt cerebral trauma. In: Proceedings IIIrd Intern Congr Neurological Surgery, Copenhagen, Intern. Congr series 110: 231–235. Excerpta Medica Foundation, Amsterdam
122. Sussman BJ, Barber JB, Goald H (1974) Experimental intracerebral hematoma. J Neurosurg 41: 177–186
123. Swansen SJ, Keller PL, Berquist TH, McLeod RA, Stephens DH (1985)

Magnetic resonance imaging of hemorrhage. AJR 145: 921–927

124. Sweet RC, Miller JD, Lipper M et al (1978) Significance of bilateral abnormalities in the CT Scan in patients with severe head injury. Neurosurgery 3: 16–21

125. Symonds CP (1940) Delayed traumatic intracerebral hemorrhage. Br Med J 1: 1048–1051

126. Teasdale G, Galbraith S, Jennett B (1980) Operate or observe ICP and the management of the "silent" traumatic intracranial hematoma. In: Shulman K et al (eds) Intracranial pressure IV. Springer, Berlin Heidelberg New York

127. Tönnis W, Friedmann G, Schmidt-Wittkamp E, Walter W (1963) Die traumatischen intrakraniellen Hämatome. Docum Geigy Ser Chir 6

128. Ugrumov VM, Zotov YY, Schenderyonok VV (1979) Early surgical treatment of traumatic hematoma and laceration foci as the main factor of favourable prognosis. Acta Neurochir (Wien) [Suppl] 28, No 1: 199–200

129. Unterharnscheidt FJ (1975) Injuries due to boxing and other sports. In: Vinken PJ, Bruyn GW (eds) Handbook of clinical neurology, Vol 26. North Holland Publ Co, Amsterdam

130. Unterharnscheidt F, Sellier K (1965) Pathomorphology of non-penetrating brain injuries. In: Proceedings IIIrd Intern Congr Neurological Surgery, Copenhagen. Intern Congr Series 110: 93–103 Excerpta Medica Foundation, Amsterdam

131. Unterharnscheidt F, Sellier K (1966) Mechanisms and pathomorphology of closed brain injuries. Conference: Head Injury Planning Committee, Chicago, 7.-9. Februar 1966. Lippincott, Philadelphia

132. Vigouroux RP, Guillermain P (1981) Posttraumatic hemispheric contusion and laceration. Progr Neurol Surg 10: 49–163. Basel, Karger

133. Weisberg LA (1979) CT and acute head trauma. Comp Tomogr 3: 15–28

134. Weisberg LA (1980) Peripheral rim enhancement in supratentorial intracerebral hematoma. Comp Tomogr 4: 145–164

135. Wilberger JE, Pang D (1981) Craniocerebral injuries from dog bite in an infant. Neurosurgery 9: 426–428

136. Woodford JE, Bogdanowicz W, Saunders RL (1974) Traumatic intracranial hematomas. Role of CSF leakage. JAMA 227: 1152–1154

137. Wozney YH, Latschaw RE, Gur D, Good W (1985) Central herniation revealed by focal decrease in blood flow without elevation of intracranial pressure. A case report. Neurosurgery 17: 641–643

138. Yashon D, Kosnik EJ (1978) Chronic intracerebral hematoma. Neurosurgery 2: 103–106

139. Yoshida M, Hayashi T, Karamoto S, Hiyoshi Y, Yokoyama T (1979) Traumatic intracranial hematomas in hemophilic children. Surg Neurol 12: 115–118

140. Young HA, Gleave JRW, Schmidek HH, Gregory S (1984) Delayed traumatic intracerebral hematoma: Report of 15 cases operatively treated. Neurosurgery 14: 22–25

141. Zilkha A (1983) Intraparenchymal fluid-blood level. A CT sign of recent intracerebral hemorrhage. J Comput Ass Tomogr 7: 310–305

142. Zimmerman RA, Bilaniuk LT, Dolinskas CA *et al* (1977) Computed tomography of acute intracerebral haemorrhagic contusion. Comp Axial Tomogr 1: 271
143. Zimmerman RA, Bilaniuk LT (1980) Computed tomography of acute intratumoral hemorrhage. Radiology 135: 355–359
144. Zimmerman RA, Deck MDF (1986) Intracranial hematomas. Imaging by High-Field MR. Radiology 159: 565
145. Zuccarello M, Tavicoli R, Pardatscher K, Cervellini P, Fiore D, Mingrino S, Gerosa M (1981) Posttraumatic intraventricular hemorrhages. Acta Neurochir (Wien) 55: 283–293
146. Zülch KJ (1968) Zur Frage der posttraumatischen Spätapoplexie. In: Alema G, Bolles G *et al* (eds) Brain and mind problems. Pensiero Scientifico Publ, Roma, pp 953–958
147. Zülch, KJ (1985) Die traumatische Spätapoplexie. Fortschr Neurol Psychiatr 53: 1–12
148. Zülch KJ (1989) Delayed posttraumatic apoplexy. Neurosurg Rev 12: [Suppl 1] 252–253

Posttraumatic Cerebellar Contusions and Hematomas

J.W. GLOWACKI

Neurotraumatology Clinic, Academy of Medicine, Cracow (Poland)

With 23 Figures

Contents

Introduction

Cerebellar contusions and hematomas result from local trauma to the occipital area, mainly from hitting a solid object with the back of the head and from occipital blows. But the protection of the cerebellum by the strong musculature of the neck, the massive bone with smooth and regular internal contours, the elastic tentorium of the cerebellum and the "fluid cushions" of the fourth ventricle and subarachnoid cisterns make up a "buffer" significantly reducing the amount of energy delivered at the moment of impact. Therefore cerebellar lesions are significantly less common than lesions of the cerebral hemispheres.

Clinical diagnosis, rather rare in the past, became somewhat more frequent only with the availability of Computed Tomography – CT – and Magnetic Resonance Imaging – MRI –. In our Series there were 116 such cases out of 9.584 treated head injuries.

Cerebellar Contusions

Incidence, Distribution of Age and Sex

Gurdjian (1971) in a collection of 152 autopsy studies in patients who died after cerebral contusion found only 18 cases, 12%, with a cerebellar contusion. Ugriumow and Zopow (1978) reported a frequency of 1% of all cerebral contusions. Only Arne (1976) reported traumatic cerebellar dysfunction in 30% of head injuries.

In our Institute in Cracow over a period of 16 years out of 2009 intracranial hematomas, we observed 116 patients referred to the clinic with an initially suspected posterior fossa hematoma. There were 84 men (75%), 6 children up to 14 years (5%), the youngest was 7 years old.

Clinical Signs and Symptoms

The varying clinical symptomatology comprises unilateral, rarely bilateral (6%) occipital bone fracture, cerebellar flaccidity, ataxia, dysmetria, dysarthria, resting and motor imbalance of various degree, rise of intracranial pressure, disturbances of consciousness, stiffness of the neck, hemorrhagic or xanthochromic CSF with normal or slightly elevated pressure, occipital headache, sometimes vomiting, nystagmus in decreasing intensity. In the subacute and chronic cases there is often lack of disturbances of vital functions, but papilloedema appears later.

In the past in some patients it happened that the symptomatology disappeared after one week spontaneously, or the exploratory operation was performed in the subacute stage 3–5 days after trauma. During the operation a suspected posterior fossa hematoma was excluded and the only pathological finding was the cerebellar contusion.

Based on our series of 55 cases we reported in 1979 about the isolated cerebellar contusion syndrome. Up to 1986 the number of contusions rose to 116, up to 1989 to 118 (Glowacki 1989). Present experience confirms our primary observations, that in isolated "pure" contusion of the cerebellum the clinical course is always favorable irrespective of the patient's age. In recent years we have observed 3 patients with both cerebellar and hemispheric contusion and secondary brain stem damage leading to death.

Diagnosis

Cerebellar contusions have until recently been recognized on the basis of a clinical syndrome. Now 10 of our patients have had CT scans, whereby the appropriate tomographic cuts must include the basal parts of the cerebellar hemisphere. Up to seven days after the trauma the result was negative in two cases, unequivocally positive in four patients with a hyperdense non-homogenous diversified focus of the "mottled" or "salt and pepper" appearance of the contused area (Figs. 1a and b). In another four cases a hypodense focus was found. Among 35 patients with hematomas and hygromas of the posterior cranial fossa, clinically latent cerebellar contusion was detected 10 times during operation.

MRI, with the elimination of bone artefacts, will improve the diagnostic possibilities in the posterior fossa. Progressive resorption of extravasations in the foci of cortical cerebellar contusions leads to similar changes as are observed in the cerebral hemispheres. Cortical atrophy of varying degrees around the focus and enlargement of the fourth and lateral ventricles occur as early as 6–8 weeks after injury.

Residual cerebellar symptoms, ranging from minor to major, should be ascribed to an atrophic process (Figs. 2 and 3).

Differential Diagnosis Among Cerebellar Hematomas and Contusions

There are only a few differences among the clinical signs and symptoms of posterior fossa hematomas and cerebellar contusions (Fig. 4). A rise of intracranial pressure occurs earlier in hematomas than in cerebellar disturbances. Rapidly progressing recovery differentiates the course of cerebellar contusions from that of hematomas. If there is deterioration, CT scanning should be repeated.

The *treatment* of cerebellar contusions corresponds to that of supratentorial contusions, as mentioned by Vigouroux and Guillermain in this volume.

Posterior Fossa Hematomas

Our series consists of 35 traumatic hematomas and hygromas out of the total of 2044 supra- and infratentorial cases (Table 1). This frequency of 1.7% is similar to the 1.2% of Cordobes *et al.* (1981). Frowein *et al.* (1989, 1990) observed 5 cases.

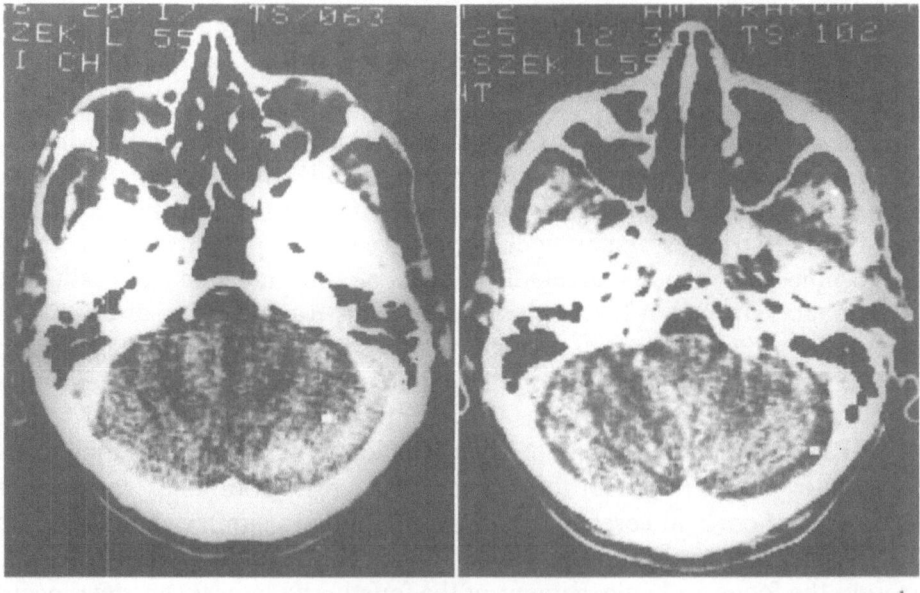

a   b

Fig. 1. a) CT scan shows contusional focus of the right cerebellar hemisphere. b) The same patient. CT scan four weeks after first examination – shows a hypodense area in the former contusional focus

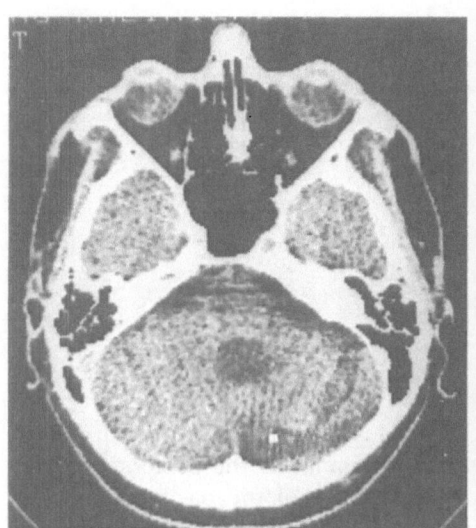

Fig. 2. In CT scan cortical atrophy of the right cerebellar hemisphere in former contusional area. Enlargement of the fourth ventricle

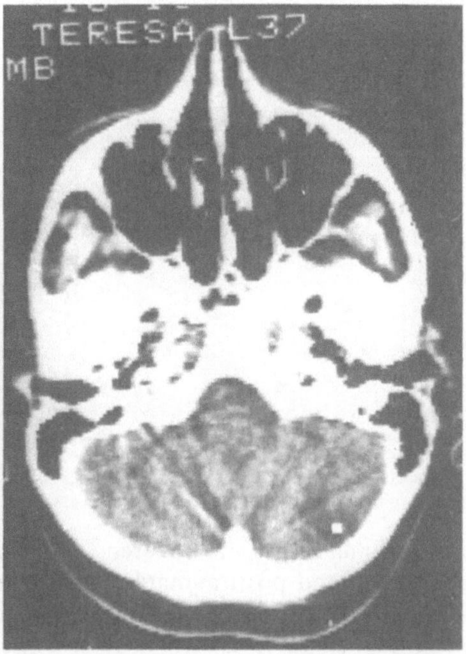

Fig. 3. CT scan two years after clinically diagnosed contusion of the cerebellum – shows a small cortical porencephalic cavity

Table 1. *Author's Series of Posttraumatic Intracranial Lesions 1971–1986*

		Supratentorial		Infratentorial	
Cerebral contusion		1468			116
Hematomas	epidural		387	20	
	subdural		1041	1	
	intracerebral		432	9	
Hygromas	subdural		149	5	
		2009	2009	35	35
Total		3477			151

Epidural Hematomas

In 1964 we reported on two cases (Glowacki 1964). In the last 16 years we have treated 20 patients: that is 5% compared to our 387 cases with infratentorial lesions (Fig. 5).

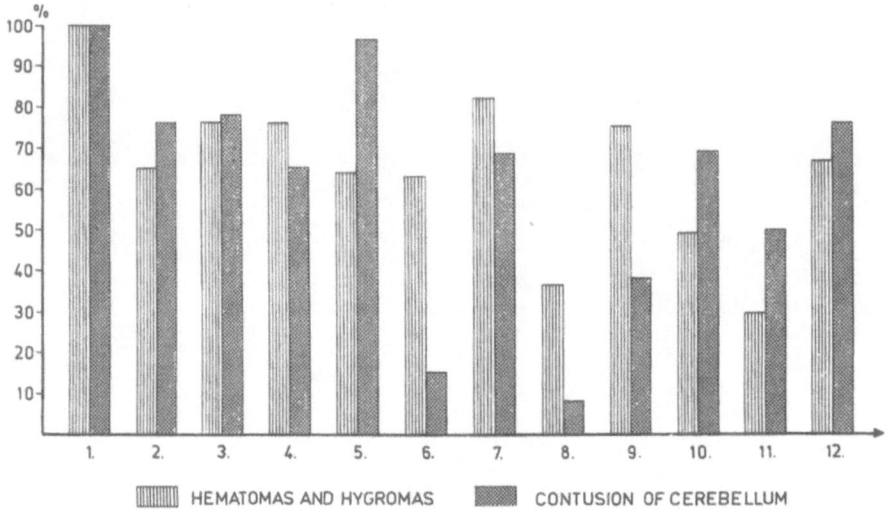

Fig. 4. Similarities and differences of signs and symptoms in the contusion of cerebellum and posttraumatic hematomas and hygromas in the posterior cranial fossa. *1*. Local trauma on occipital region, *2*. fracture of the occipital bone, *3*. meningeal signs, *4*. disturbances of consciousness, *5*. erythro-or-xanthochromia of CSF, *6*. CSF pressure over 20 mm H_2O, *7*. headache, nausea, vomiting, *8*. papilloedema, *9*. impairment of cranial nerves, *10*. nystagmus, *11*. disturbances of equilibrium, *12*. other cerebellar signs and symptoms

SIGNS & SYMPTOMS	CASES NUMBER TOTAL	1	2	3	4	5	6	7	8	9	10	11	12	13	14	15	16	17	18	19	20
SEX M	17	+	+	+	+		+	+	+	+	+	+	+	+		+	+	+		+	+
SEX W	3				+										+			+			
LOCAL TRAUMA OF THE OCCIPITAL REGION	20	+	+	+	+	+	+	+	+	+	+	+	+	+	+	+	+	+	+	+	+
FRACTURE OF THE OCCIPITAL BONE	13	+		+	+	+	+	+		+		+	+		+		+	+			+
NECK STIFFNESS	12	+	+	+	+	+	+	+				+	+			+	+	+			
LOSS OF CONSCIOUSNESS	13	+		+	+	+			+			+	+		+		+	+	+	+	+
ERYTHRO-OR-XANTHOCHROMIA OF CSF	4	+							+					+		+					
HEADACHE, NAUSEA, VOMITING	18	+	+	+	+	+	+	+	+	+	+	+	+	+	+		+	+	+	+	
PAPILLEDEMA	9	+	+	+						+	+	+				+			+	+	
IMPAIRMENT OF CRANIAL NERVES	14	+	+		+		+	+	+	+	+	+	+		+		+	+	+		
NYSTAGMUS	6	+	+			+		+		+	+										
DISTURBANCES OF EQUILIBRIUM	6	+	+			+	+			+		+									
ATAXIA	5	+				+				+	+	+									
BRADYCARDIA	2				+													+			
RESPIRATORY DISTURBANCES	4				+		+											+			+
OPERATIVE TREATMENT	15	+	+	+	+	+	+	+	+	+	+	+	+		+		+	+	+		
DECEASED	4								+	+					+						+
SURVIVAL	16	+	+	+	+	+	+	+			+	+	+		+	+	+	+	+	+	+

Fig. 5. Signs and symptoms occurring in 20 cases of posterior fossa epidural hematoma

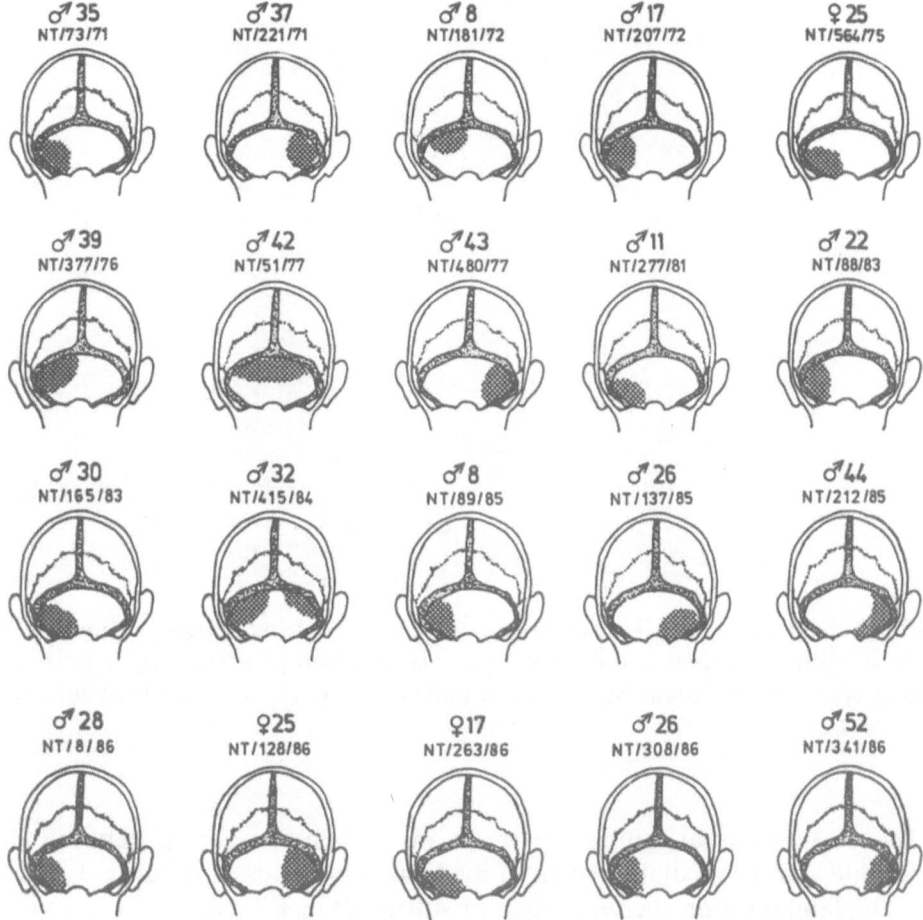

Fig. 6. Sex, age and localisation in 20 cases of posterior fossa epidural hematoma

Age and Sex Distribution; Location

Only three of our 20 patients were under 18 years old (8, 8 and 17 years), but more young children with head injuries have been treated predominantly on a neurosurgical pediatric ward. Figure 6 shows the sex and age of the patients and the location in our 20 cases.

Etiology

The local impact to the occipital region usually results in contusion and laceration of the soft tissue (Fig. 7). Bone deformation causes separation of the dura mater, detaching the sinus and small superficial venous vessels.

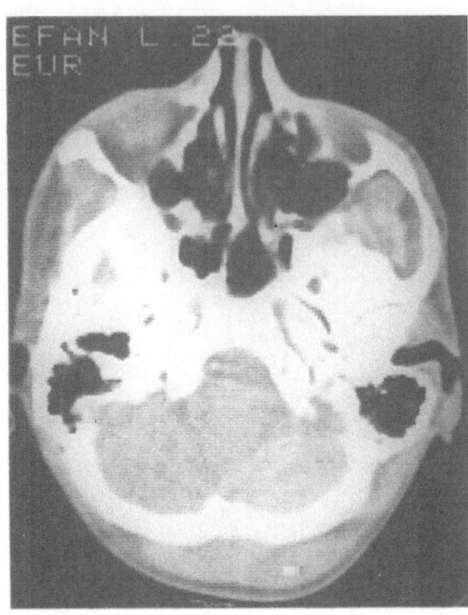

Fig. 7. Non contrast CT scan shows an extensive hemorrhagic imbibition of the occipital soft tissues and only slight visualisation of a large right posterior fossa epidural hematoma. See: the same patient in Fig. 11 – after contrast enhancement

The postero-lateral injury can result in a fracture running between the midline and the medial segment of the transverse sinus or produce diastasis of the lambdoid or occipito-mastoid sutures (Fig. 8 a–c).

In our series of epidural hematomas of the posterior fossa fractures have been shown radiologically in 64%, by Roda *et al.* (1983) in 67%. In six of our cases we could not identify the source of the bleeding. There were bilateral hematomas in two patients.

Pathological Effect of Hematoma on the Posterior Fossa Structures

A collection of 15 to 30 ml of blood within the subtentorial space results in a severe functional disturbance. Obstructive hydrocephalus and invagination of cerebellar tissue into the foramen magnum and the incisura tentorii may develop.

We regard as acute those cases operated on within 24 hours after injury, as subacute when signs evolve within 2 to 14 days after a lucid interval, and as chronic when signs evolve after more than 15 days.

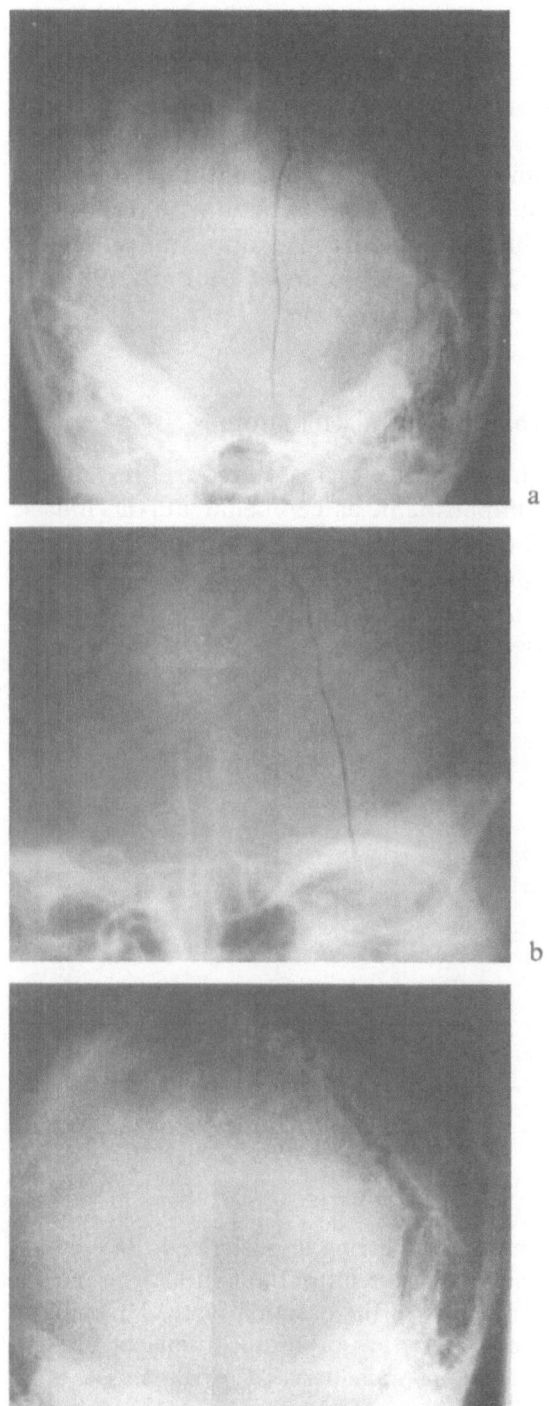

Fig. 8a–c. Suboccipital fractures.
Explanation in text, page 182

Acute Posterior Fossa Epidural Hematoma

In our series acute hematomas were present in 6 patients (30%), subacute in 8 patients (40%) and chronic in 6 cases. Routine use of the CT scan enabling earlier diagnosis may change these proportions. In rapidly developing – within 6 hours after trauma – epidural hematomas a lucid interval may be missing, and signs of acute brain stem compression with respiratory disturbances appear (Guillermain 1981) before the appearance of any localizing signs (Schneider *et al.* 1951, Teasdale and Galbraith 1981). In these hyperacute forms the outcome is usually rapidly fatal.

Subacute and Chronic Epidural Hematomas

Patients with subacute and chronic posterior fossa epidural hematoma usually show signs of cerebellar impairment, as cerebellar ataxia, muscle hypotonia of the leg homolateral to the hematoma, bulbar and long tract signs, progressing steadily or intermittently. Usually signs of increased intracranial pressure evolve, such as drowsiness, headache, slowness of movement and apathy, papilloedema and restlessness (Fig. 9).

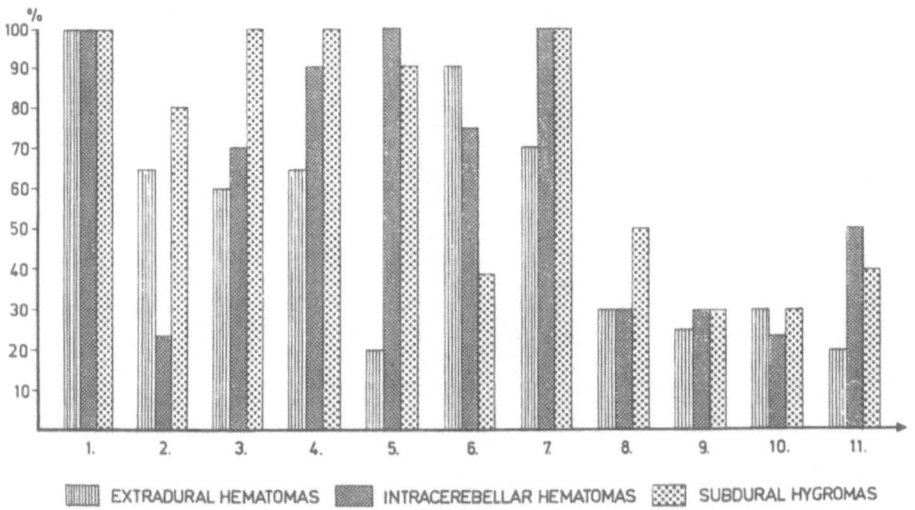

Fig. 9. Comparison of signs and symptoms occurring in posterior fossa epidural hematoma, intracerebellar hematomas and subtentorial subdural hygroma: *1.* Local trauma on occipital region, *2.* fracture of the occipital bone, *3.* meningeal signs, *4.* disturbances of consciousness, *5.* erythro-or-anthochromia of CSF, *6.* headache, nausea, vomiting, *7.* impairment of cranial nerves, *8.* nystagmus, *9.* other cerebellar signs and symptoms, *10.* disturbances of equilibrium, *11.* bradycardia, respiratory disturbances

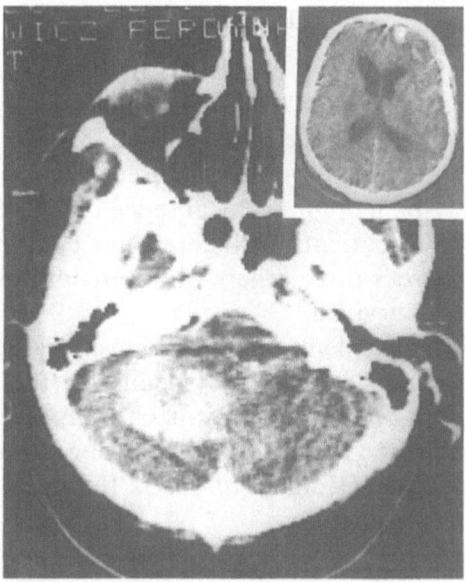

Fig. 10. Left intracerebellar hematoma associated with contre-coup lesion of the right frontal lobe

Cases with an interval of weeks and even months have become exceptional since the availability of CT (Figs. 11 to 16).

Accompanying Lesions

The occipital impact could simultaneously injure structures in the supratentorial compartment by direct or contre-coup mechanism as in Fig. 10. Jennett and Teasdale (1981), Reigh and O'Connell (1962) in a survey of 80 posterior fossa epidural hematoma cases, found 17 accompanying supra-tentorial hematomas. Figure 10 shows a left intracerebellar hematoma and a contre-coup lesion of the right frontal lobe.

Clivus Extradural Hematoma

This extremely rare entity was described by Orrison *et al.* (1986) in a 8.5 year-old boy who had been hit by a motorcycle causing multiple injuries. The patient died in deep coma. At autopsy there was dislocation of the left atlanto-occipital joint and of the odontoid process of the axis. Under the dura, separated over the clivus from the foramen magnum up to the dorsum sellae, the epidural hematoma was one cm thick. Retrospectively a high density collection along the clivus on the CT scan was seen.

Diagnosis

If no CT scan or MRI are available the clinical symptoms already mentioned have to be considered carefully. In plain X-rays, occipital projections included, fracture of the occipital bone yields significant information. Ventriculography is practically no longer used.

Angiography may be helpful in the absence of CT. But Cordobes *et al.* (1981) report that three patients with cerebellar epidural hematoma died after CT scanning whereas three others who were immediately operated on without CT scanning, survived.

In subacute and chronic hematomas the density values of the hematoma on the CT scan decrease and become isodense in the surrounding brain area, difficult to be seen on the native scan. Contrast enhancement appears in this stage as a ring around the hematoma reaching a peak about 10 minutes after injection as in Fig. 16 (Roda *et al.* 1983). If the rim is present without contrast it indicates calcification.

Treatment

As a rule a posterior fossa epidural hematoma detected by CT scanning should be evacuated immediately.

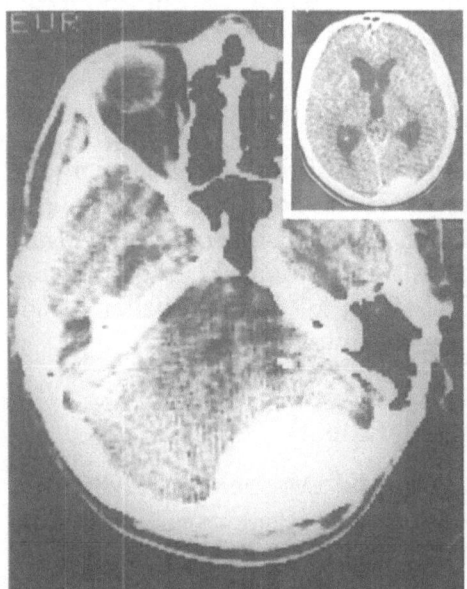
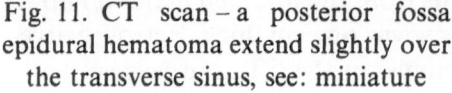
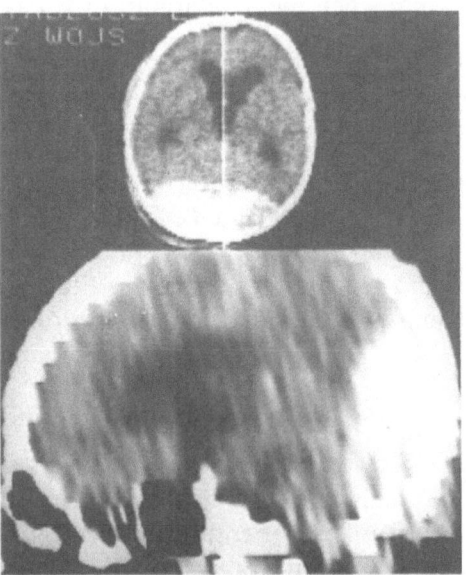

Fig. 11. CT scan – a posterior fossa epidural hematoma extend slightly over the transverse sinus, see: miniature

Fig. 12. Large bilateral occipital epidural hematoma extending into the posterior fossa

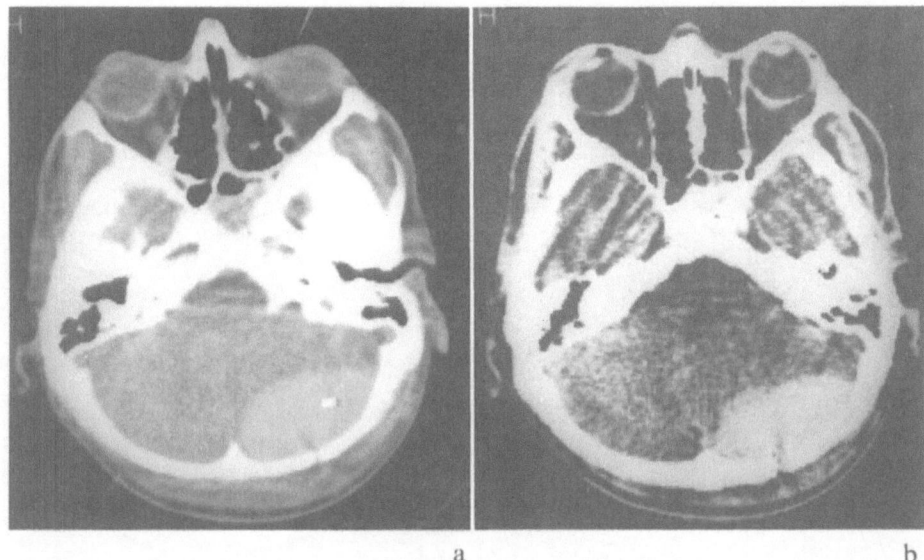

a   b

Fig. 13. A typical, large, lenticular epidural hematoma located on the right side of
the posterior fossa. a) In non contrast scan. b) Same with contrast enhancement
within the hematoma. Note also the fracture of the occipital squama

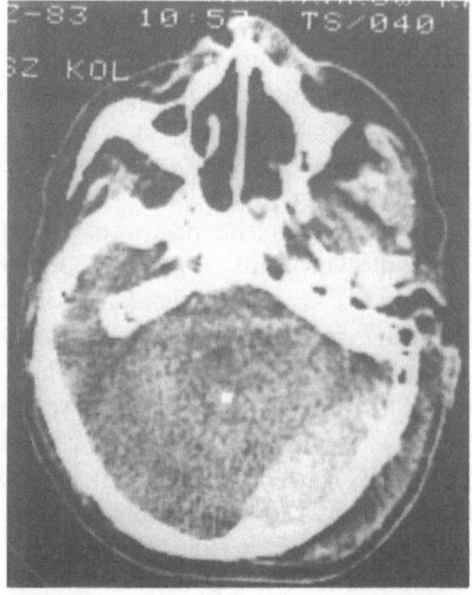

Fig. 14. A spindle-shaped mass of the blood density located in the right posterior
fossa. Note some decrease in density clot. CT scan 8 days after occipital trauma

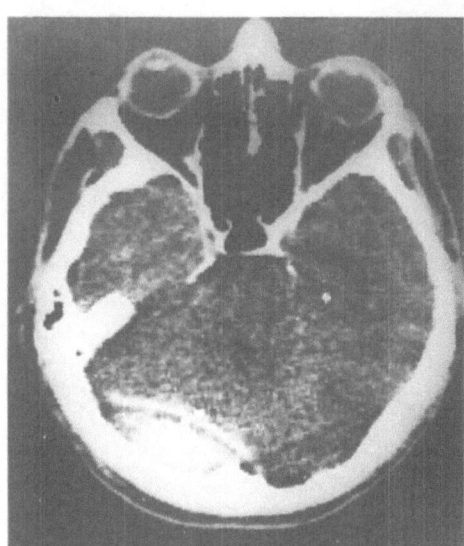 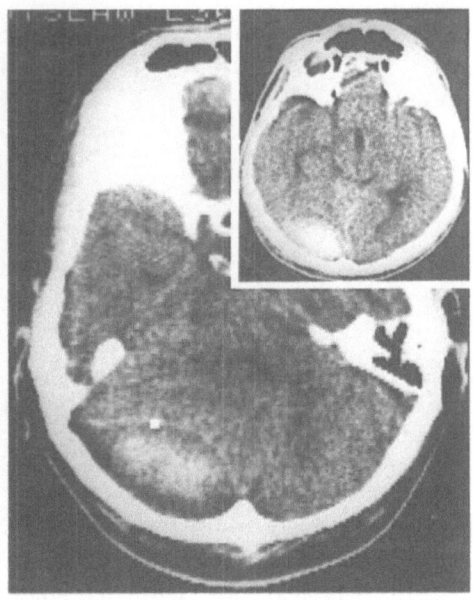

Fig. 15. Chronic posterior fossa epidural hematoma compressing left cerebellar hemisphere. CT scan shows a biconvex area with rim enhancement

Fig. 16. A large chronic posterior fossa epidural hematoma with small extension over the left transverse sinus. Note the beginning of the decrease of density and rim enhancement

If none of the contemporary diagnostic facilities are available and the patient's state is very urgent, exploratory burr holes uni- or bilaterally over the cerebellar hemispheres, if necessary bilateral subtentorial craniectomy, are advisable. One must never delay the operation. Loss of consciousness before operation indicates a poor prognosis.

A special problem arises from the non-operative treatment of fully conscious patients, without symptoms of raised ICP and without signs of brain stem injury, in whom CT scanning revealed an infratentorial epidural hematoma of small volume, sometimes with radiological signs of resorption. In four patients aged 17, 25, 26 and 44 years of the author's series such symptoms were found. All of them were admitted within 3–4 weeks after the injury. They were conscious (GCS 14, 11, 14, 15), with mild cerebellar symptoms, headaches and CT evidence of small hematoma and signs of resorption. Based on numerous reports concerning resorption of epidural hematoma (Illingworth and Shawdon 1983; Pang *et al.* 1983; Weaver *et al.* 1981) and on the author's experience, gained in the medical treatment of multiple intra-cerebral hematomas, they were not referred for operation, but were kept under constant supervision, with all the necessary equipment

at hand to deal with possible complications. Medical treatment was in-
stituted including steroids, osmotic agents reducing cerebral energy metabolism and cerebral blood flow. Control CT scans revealed gradual resorption
correlating with progressive improvement of the neurological state of the
patients.

The above observations are in accordance with the correct opinion of
McLaurin (1986), expressed in the epilogue of the first volume of "Advances
in Neurotraumatology" (p. 249), namely "this mode of management of
epidural hematoma should be utilized only under exceptional and well
controlled circumstances". Indeed, only strict selection of exceptional cases
authorizes the abandonment of surgery and the utilization of medical
treatment. These exceptional cases include hematomas found on routine
CT scanning, which are of small volume and which do not produce
intracranial hypertension. Our four patients fulfilled exactly these criteria.

Mortality

With CT scanning the mortality rate was reduced. In the author's recent
series of 20 cases there were 4 deaths (20%). The high mortality rate of 50%
observed by Tsai *et al.* (1980) in their series of 14 cases despite the use of CT
may be related to an aggressive policy in scanning even moribund patients
(John *et al.* 1986). However 7 cases of Garza-Mercado (1983) may show
that even patients with severe bulbar signs, if they are acutely ill or
moribund, can be saved if the interval between trauma and the operation is
short (Table 2).

Posterior Fossa Subdural Hematomas

Incidence

Subdural hematomas of the posterior cranial fossa are less frequent
than epidural hematomas from which they do not differ much in clinical
course. Lang and Reding (1985) found 21 cases in an autopsy series of 737
subdural hematomas, *i.e.* 3%, before the advent of CT scanning. In our
series of 1041 subdural hematomas treated in our clinic in Cracow, there
was only one "true" subtentorial subdural hematoma and another one
coexisting with an epidural hematoma (Table 3).

Pathology

The mechanism of subdural subtentorial hematoma formation is similar
to that of a subdural supratentorial hematoma: Bleeding associated with
tearing of veins from the cerebellar surface. Or the subdural hematoma

Table 2. *Mortality in Posterior Fossa Epidural Hematomas*

Author	Year of publication	Number of cases	Death	Author	Year of publication	Number of cases	Death
Hooper	1954	7	3	Tsai et al.	1980	14	7
Fisher	1958	8	1	Roda et al.	1983	3	0
Wright	1966	6	2	Garza-Mercado	1983	7	0
Jamieson, Yelland	1968	12	4	John et al.	1986	4	0
(after Jamieson)	/1976/			Guillermain	1986	10	4
Stone et al.	1979	4	0				

Table 3. *Incidence of Subtentorial Subdural Hematomas*

Author	Year of publication	Number of cases	
		Subdural subtentorial hematoma	Subdural hematomas observed
Munro	1934	1	62
(cit Wright)	/1966/		
Mc Kissock	1960	2	389
Ciembroniewicz	1965	3	539
Wright	1966	5	361
Jamieson, Yelland	1972	7	553
Miles, Medley	1974	2	418
Lang, Reding	1985	21*	737

* Autopsy material.

results from contusion and laceration of a cerebellar hemisphere (Jamieson 1976). Also combined subdural and intracerebellar hematomas were reported (Clitherrow *et al.* 1969).

Diagnosis

On the basis of clinical symptoms the diagnosis of a subdural cerebellar hematoma may only be tentative because there are no specific symptoms. Our own single case occurred in a 22-year-old man who sustained a head injury with an occipital blow with bone fracture and with loss of consciousness.

As with the epidural hematomas, CT scan demonstrates acute subdural bleeding as a hyperdense collection, but the differentiation between epidural and subdural hematoma can be difficult.

Treatment

In pressure-threatening states a burr hole over the cerebellar hemisphere should be used as a decisive procedure. Known hematomas must be removed through uni- or bilateral craniectomy, incision of the dura and washing out the hematoma. Exact identification of the hemorrhagic site and termination of bleeding is mandatory.

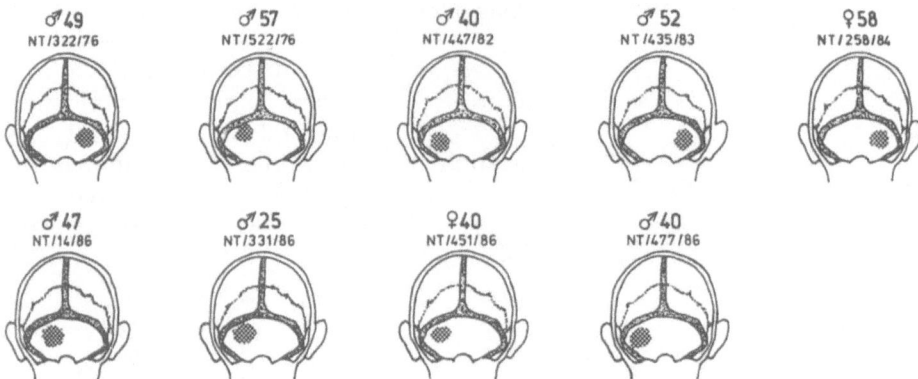

Fig. 17. Sex, age and location in 9 cases of intracerebellar hematoma

Mortality

Due to the rapid increase in intracranial pressure and brain stem compression, the mortality is high, particularly in the acute stage. In a series (collective study) of 20 cases of Ciembroniewicz (1965), of the 16 patients operated upon 4 died. Our two patients operated on in the subacute stage survived.

Intracerebellar Hematomas

Incidence

Infratentorial intracerebellar hematoma are rare, Lang-Reding (1985) reported recently about 5 cases, Tsai *et al.* (1980) 12 patients, Firsching *et al.* (1990) 5 cases; our own series consist of 9 cases (Fig. 17).

The natural history of traumatic intracerebral and intracerebellar hematoma is similar. CT investigation and operation indicate that an intracerebellar hematoma is usually associated with a cortical contusion. Mc Laurin *et al.* (1989) draw attention to intracerebral hematomas in the battered infant.

Signs, Symptoms, Diagnosis

Wright (1966) emphasized diagnostic difficulties mainly because of the paucity of symptoms. X-ray investigation frequently reveals fracture of the occipital bone, in 3 of our 9 cases (Fig. 18). With the availability of CT (Fig. 19) the frequency of diagnosis has improved. Tibbs *et al.* (1981) emphasize that patients with cerebellar contusion revealed by CT are in potential risk of delayed intracerebellar hematoma, which we observed

SIGNS & SYMPTOMS		CASES NUMBER TOTAL	1	2	3	4	5	6	7	8	9
SEX	M	8	+	+	+	+		+	+	+	+
	W	1					+				
LOCAL TRAUMA OF THE OCCIPITAL REGION		9	+	+	+	+	+	+	+	+	+
FRACTURE OF THE OCCIPITAL BONE		3	+					+		+	
NECK STIFFNESS		5		+	+			+	+	+	
LOSS OF CONSCIOUSNESS		7	+	+	+		+	+	+	+	
ERYTHRO-OR-XANTHOCHROMIA OF CSF		5		+	+		+		+	+	
HEADACHE, NAUSEA, VOMITING		6		+	+	+		+	+	+	
PAPILLEDEMA		1				+					
IMPAIRMENT OF CRANIAL NERVES		6		+	+	+	+	+	+		
NYSTAGMUS		3		+	+			+			
DISTURBANCES OF EQUILIBRIUM		1				+					
ATAXIA		2		+		+					
BRADYCARDIA		3	+				+	+			
RESPIRATORY DISTURBANCES		4	+				+	+	+		
OPERATIVE TREATMENT		6	+	+		+	+		+	+	
DECEASED		4	+				+		+	+	
SURVIVAL		5		+	+	+		+			+

Fig. 18. Signs and symptoms in 9 cases of intracerebellar hematoma

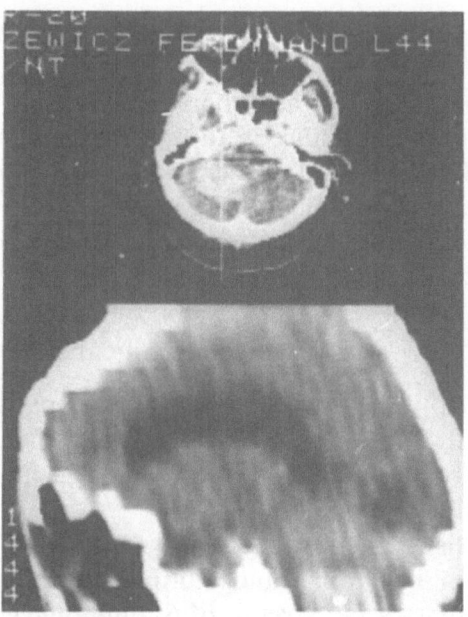

Fig. 19. Hematoma of the left cerebellar hemisphere

twice. Therefore repeated CT is recommended if there is the slightest deterioration of the neurological state.

We used a diagnostic trephine opening in one case, ventriculography in one case and CT scanning in 7 cases to detect the hematoma. Sometimes co-existence of a supratentorial frontal contusion and intracerebral hematoma opposite to the cerebellar hematoma is found as in Fig. 10.

Surgical Treatment

In my experience, with CT scan well-located intracerebellar hematomas of sufficient volume should be removed surgically, in particular those that cause displacement of the posterior fossa structures, brain stem compression, obstructive hydrocephalus and raised intracranial pressure. In emergency cases it is useful to begin the operation with ventricular drainage.

In the author's series five patients were operated on, usually in the prone position by means of a typical opening of the posterior fossa. In two patients with a chronic hematoma of small volume demonstrated by CT scanning, the operation was abandoned. Firsching et al. (1987) distinguish sole ventricular drainage, sole direct evacuation of the hematoma, or both together with the drainage before or after the evacuation of the hematoma. Because of the limited number of 5 cases and because of individual circumstances and a different interval between trauma and operation in each case, no bias for one or the other of the methods is possible.

Results

Three out of our 16 cases died after the operation within the first 24 hours after injury and admittance in the state of severe distress and brain stem impairment. Evacuation of the hematoma and decompression did not improve their state. The mortality rate was 44% and thus is similar to that found by Wright, 1966. Both our patients who survived were operated on in the subacute state; they were conscious.

Conservative Treatment

Even hematomas of large volume may show spontaneous resorption, controllable by CT (Pozzati et al. 1981), leading to the disappearance of clinical symptoms, or, depending on the extent of the injury, to the maintenance of the residual symptoms.

Among our patients there was one in whom the CT scan revealed an acute intracerebellar hematoma and multiple supratentorial hematomas. Because of deep coma, a GCS score of 6, pulmonary complications and circulatory failure he did not qualify for the operation. For 12 days he was

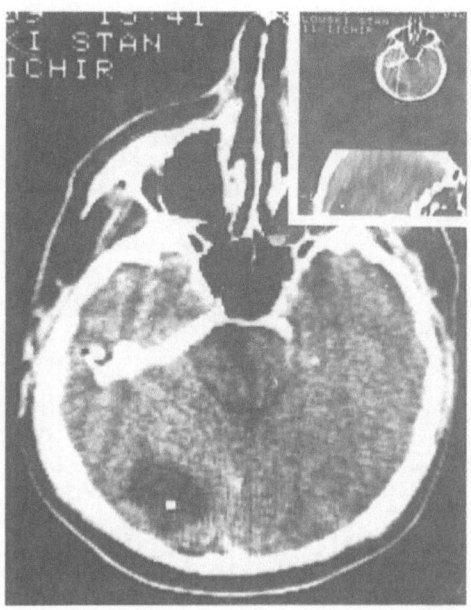

Fig. 20. Posttraumatic porencephalic cyst in the left cerebellar hemisphere

on a respirator and was given steroids and osmotic agents. After improvement CT studies demonstrated advanced resorption of the cerebellar and supratentorial hematomas. After 8 weeks he was discharged from the clinic with resolving neurological symptoms. In these cases later CT scans may detect porencephalic cysts as a persistent trace of earlier intracerebellar hematoma (Figs. 20, 21a and b).

Subtentorial Subdural Hygroma

Incidence

Fisher *et al.* (1958) and Jamieson (1976) each have seen seven patients with a subtentorial subdural hygroma; in our series there were five patients: Acute hygroma was found twice, subacute also twice and a chronic hygroma, operated after 14 days, was observed in one patient.

Signs, Symptoms, Diagnosis

The symptoms, observed in our series, are summarized in Figs. 22 and 23.

As with subdural supratentorial hygromas (Glowacki 1963, Matsumoto and Tamaki, 1986) there are no specific symptoms differentiating subdural hygroma from other traumatic complications in the posterior fossa.

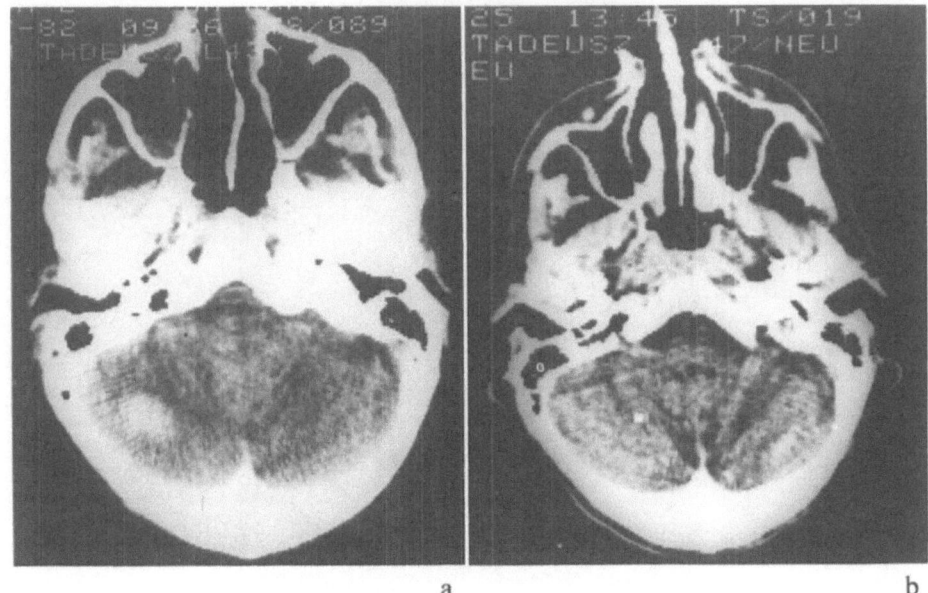

Fig. 21. a. CT scan reveals the left intracerebellar hematoma. b. The same patient. Central CT scan 3 months after the first examination. Total resorption of the hematoma

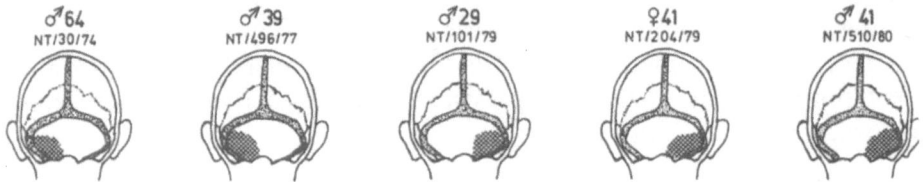

Fig. 22. Sex, age and location of subtentorial subdural hygroma

Masuzawa *et al.* (1983) reports on clear distinction by CT scan between hematomas and hygromas: When compared to the density of the ventricular cavity, all of the low density hematomas were higher in density, while all hygromas appeared CSF dense or lower.

With MRI subdural collections composing hygromas can be differentiated from mixed subdural hematomas, but MRI is superior to CT scanning only three days after trauma. In the acute phase CT remains a decisive diagnostic method.

In our studies all patients have had "pure" subdural hygromas. Langfitt *et al.* (1961) and Garza-Mercado (1983) observed subdural cerebellar hygromas accompanied with an extradural hematoma of the posterior cranial fossa.

SIGNS & SYMPTOMS	Cases Number (per)	1	2	3	4	5
SEX M	4	+	+	+		+
SEX W	1				+	
LOCAL TRAUMA OF THE OCCIPITAL REGION	5	+	+	+	+	+
FRACTURE OF THE OCCIPITAL BONE	4	+	+	+		+
NECK STIFFNESS	4	+	+		+	+
LOSS OF CONSCIOUSNESS	5	+	+	+	+	+
ERYTHRO-OR-XANTHOCHROMIA OF CSF	4	+	+	+		+
HEADACHE, NAUSEA, VOMITING	3	+		+	+	
PAPILLEDEMA	2	+		+		
IMPAIRMENT OF CRANIAL NERVES	3	+		+	+	
NYSTAGMUS	3	+			+	+
DISTURBANCES OF EQUILIBRIUM	3			+	+	+
ATAXIA	3			+	+	+
BRADYCARDIA	2	+	+			
RESPIRATORY DISTURBANCES	1		+			
OPERATIVE TREATMENT	5	+	+	+	+	+
DECEASED	1		+			
SURVIVAL	4	+		+	+	+

Fig. 23. Signs and symptoms in 9 cases of subtentorial subdural hygroma

Treatment

The finding of a large hygroma requires surgical intervention. In our series all cases have been revealed by a diagnostic trepanation which was then changed into a therapeutic procedure. Reaccumulation of fluid in the cyst, occasionally found in supratentorial hygromas, has not been seen in the posterior cranial fossa.

The CT detection of a subdural hygroma of small volume in patients with mild cerebellar disorders, without symptoms of increased intracranial pressure and with gradual improvement, may be the reason for refraining from surgical intervention, under close observation and control CT studies, which may show progressive resorption of the fluid (Bakay *et al.* 1980, French *et al.* 1978).

Results

The seven non-complicated cases of Fisher *et al.* (1958) and also four out of our five patients recovered. In one of our cases who died, the operation was performed several hours after severe trauma which had produced severe disturbances of consciousness, the GCS was 6. The hygroma of 30 ml volume with a slight admixture of blood, coexisted with tetraplegia after an injury to the cervical spine at C 1 level, what makes pre-existing subarachnoidal cyst in the posterior cranial fossa very likely, as it is known with supratentorial cysts (Kulali and v. Wild 1989). Death resulted from circulatory and respiratory failure.

Depressed Fractures and Tearing of Large Sinuses of the Posterior Fossa

Occipital squama fractures with fragmentation and depression of bone fragments into the posterior cranial fossa and sinuses are uncommon. Those bone fragments may plug the torn sinus. Removal of the bone may be followed by sudden bleeding, requiring very close cooperation with the anaesthetist. A sufficiently wide craniectomy around the depressed area permitting hemostasis is mandatory.

Another serious complication of depressed fractures of the posterior fossa may be a syndrome of acute central cervical spine injury (Motozaki and Yamamoto 1989).

How to Do It

No clinical symptoms were observed in any of our patients which could unequivocally indicate the presence of a hematoma in the posterior fossa. It was suspected only because the trauma was localized in the occipital region and because a fracture of the occipital bone was seen in almost all the cases (Glowacki et al. 1989). Today the CT or MRI reveal the hematoma. As mentioned before, as a rule an extracerebellar hematoma has to be evacuated immediately. In emergency cases a ventricular drainage may be useful at the beginning of the operation. Exceptionally intracerebral cerebellar hematomas may be treated conservatively or by ventricular drainage only, under supervision and repeated control CT scan.

References

1. Arne L (1976) Posttraumatic cerebellar signs and symptoms. In: Vinken PJ, Bruyn GW (eds) Injuries of the brain and skull (Part II). Handbook of clinical neurology, Vol 23. North-Holland, Amsterdam, pp 459–463
2. Bakay L, Glasauer FR, Alker GJ Jr (1980) Head injury. Little Brown, Boston
3. Ciembroniewicz JE (1965) Subdural hematoma of the posterior fossa. Review of the literature with addition of three cases. J Neurosurg 22: 354–373
4. Clitherow NR, Fowler Ch, Sedzimir CB (1969) Combined intracerebellar and posterior fossa subdural hematoma – Case report. J Neurosurg 30: 744–746
5. Cordobes F, Lobato RD, Rivas JJ et al (1981) Observations on 82 patients with extradural hematoma. Comparison of results before and after the advent of computerized tomography. J Neurosurg 54: 179–186

6. Firsching R, Frowein RA, Thun F (1987) Intracerebellar hematomas; eleven traumatic and non-traumatic cases and a review of the literature. Neurochirurgia 30: 182–185
7. Firsching R, Huber M, Frowein RA (1991) Cerebellar hemorrhage: Management and prognosis. Neurosurg Rev, in press
8. Fisher RG, Kim JK, Sachs E Jr (1958) Complications in posterior fossa due to occipital trauma – their operability. Jama 167: 176–182
9. French BN, Cobb CA, Corkill G, Youmans JR (1978) Delayed evolution of posttraumatic subdural hygroma. Surg Neurol 9: 145–148
10. Frowein RA (1990) Personal communication
11. Frowein RA, Schiltz F, Stammler U, (1989) Early posttraumatic intracranial hematoma. Advances in Neurotraumatology. Neurosurg Rev 12 [Suppl 1] 184–187
12. Garza-Mercado R (1983) Extradural hematoma of the posterior cranial fossa – Report of seven cases. J Neurosurg 59: 664–672
13. Glowacki JW (1963) Posttraumatic subdural hygromas. Pol Med Sci Hist Bull (Chicago) 5: 135–139
14. Glowacki J (1964) Two cases of epidural hematoma in the posterior cranial fossa (pol). Proc VI Congr of Polish Neurolog Neurosurg Lódz, Polfa, pp 216–217
15. Glowacki JW (1989) Posterior fossa hematoma and the contusion of the cerebellum. 9th Int Congr Neurol Surg. Book of abstracts, p 259. New Delhi, India
16. Glowacki JW, Kusmiderski J, Kwiatkowski St, Moskala M (1989) Atypical localisation of traumatic extradural hematomas. Neurosurg Rev 12 [Suppl 1]: 190–195
17. Guillermain P (1986) Traumatic extradural hematomas. Advances Neurotraumatology 1: 1–50
18. Gurdjian ES (1971) Mechanisms of impact injury of the head. Churchill Livingstone, Edinburgh, London, pp 17–22
19. Illingworth R, Shawdon H (1983) Conservative management of intracranial extradural hematoma presenting late. J Neurol Neurosurg Psychiatry 46: 558–560
20. Jamieson KG (1976) Posterior fossa hematoma. In: Vinken PJ, Bruyn GW (eds) Injuries of the brain and skull, Part II. Handbook of clinical neurology, Vol 24. North-Holland, Amsterdam, pp 343–350
21. Jennet B, Teasdale G (1981) Management of head injuries. F. R. Davis Company, Philadelphia
22. John JN St, French BN (1986) Traumatic hematomas of the posterior fossa. A clinicopathological spectrum. Surg Neurol 25: 457–466
23. Kulali A, von Wild K (1989) Posttraumatic subdural hygroma as a complication of arachnoid cysts of the middle fossa. Advances in neurotraumatology. Neurosurg Rev 12 [Suppl 1]: 508–513
24. Lang G, Reding R (1985) Schädel-Hirn- und Mehrfachverletzungen. JA Barth, Leipzig

25. Langfitt TW, Mc Queen JD (1961) Extradural hematoma of the posterior fossa with an associated space-occupying collection of spinal fluid. Case report. J Neurosurg 18: 531–534
26. Masuzawa H, Sato J, Kamitani H, Yamashita M (1983) The contents of chronic subdural hematoma and its CT density, with special reference to differentiation from subdural hygroma. Neurol Med Chir (Tokyo) 23: 123–130
27. Matsumoto S, Tamaki N (1986) Subdural hygromas. In: Vigouroux RP (ed) Advances in neurotraumatology, Vol 1. Springer, Wien New York, pp 157–172
28. McLaurin RL (1986) Extracerebral collections. Advances in neurotraumatology, Vol 1. Springer, Wien New York
29. McLaurin RL, Babcock DS, Kaufman R (1989) Traumatic intracerebral hematomas in infancy. Neurosurg Rev 12 [suppl 1]: 219–224
30. Motozaki T, Yamamoto T (1989) Unusual case of depressed fracture in the posterior cranial fossa associated with the syndrome of actue central cervical spinal cord injury. Advances in neurotraumatology. Neurosurg Rev 12 [Suppl 1]: 595–599
30a. Orrison WW, Rogdes S, Kinard RE, Williams JE, Torvik A, Sackett JF, Ammundsen P (1986) Clivus epidural hematoma: case report. Neurosurgery 18: 194–196
31. Pang D, Horten JA, Herren JM, Wilberger JE Jr, Uries JK (1983) Non surgical management of extradural hematomas in children. J Neurosurg 59: 958–971
32. Pozzati E, Piazza GC, Padovani R, Gaist G (1981) Benign traumatic intracerebellar hematoma. Neurosurgery 8: 102–103
33. Reigh EE, O'Connell TJ (1962) Extradural hematoma of the posterior fossa with concomitant supratentorial subdural hematoma with secondary hydrocephalus. Report of case and review of the literature. J Neurosurg 19: 359–364
34. Roda JM, Giménez D, Higueras AP, Blázquez MG, Alvarez MP (1983) Posterior fossa epidural hematomas: a review and synthesis. Surg Neurol 19: 419–424
35. Schneider RC, Kahn EA, Crosby EC (1951) Extradural hematoma of the posterior fossa. Neurology (Minneap) 1: 386–393
36. Teasdale G, Galbraith S (1981) Acute traumatic intracranial hematomas. Progr Neurol Surg 10: 252–290. Karger, Basel
37. Tibbs Ph A, Goldstein SJ, Smithson JR (1981) Delayed traumatic intracerebellar hematoma. Surg Neurol 16: 309–311
38. Tsai FY, Teal JS, Itabashi HH, Huprich JE, Hieshima GB, Segall HD (1980) Computed tomography of posterior fossa trauma. Comput Axial Tom 4: 291–305
39. Ugriumow VM, Zotow UV (1978) Clinic, diagnostics and treatment of severe closed cranial injury (russ.). In: Arutjunow AI (ed) Textbook of neurotraumatology. Medicina, Moskwa, pp 276–304
40. Weaver D, Pobereskin NL, Jane JA (1981) Spontaneous resolution of epidural hematomas. Report of two cases. J Neurosurg 54: 248–251
41. Wright RL (1966) Traumatic hematomas of the posterior cranial fossa. J Neurosurg 25: 402–409

Early Dynamic Evolution of Cerebral Contusions and Lacerations
Clinical and Radiological Findings

R.A. Frowein, U. Stammler, R. Firsching, G. Friedmann, and F. Thun

Neurosurgical Department and Radiological Department of the University of Cologne (Federal Republic of Germany)

With 19 Figures

Contents

Definition

Contusions of brain tissue are focal lesions of the brain parenchyma, caused by kinetic energy absorbed by the brain. *Laceration* is an injury similarly inflicted by a force but resulting in a tear of brain tissue. It is usually associated with fractures and with penetrating wounds from sharp objects (Gennarelli 1987).

The pia-arachnoid is intact over contusions but torn in lacerations (Rowbotham 1942, 1964, Graham, Adams and Gennarelli 1987).

The historical development of these definitions and the clinical correlation with commotion, contusion and concussion are discussed in detail in the chapters by Nakamura, Oprescu, Vigouroux and Guillermain in this volume.

The correlation of the histopathological and clinical findings turned out to be less close than the early concept of commotion/concussion and contusion had suggested (Ommaya *et al.* 1964, 1982). Therefore these terms are being used less frequently in recent times to denote a posttraumatic clinical syndrome. Today the diagnosis "contusion" is almost entirely reserved for lesions visualized by radiological means.

Radiological Findings

As the correlation of computerized tomography (CT) with gross pathological findings is virtually complete, this investigation has taken a key role in the management of head injuries. While there are some reports (Cooper 1982, Langfitt *et al.* 1986, Levin *et al.* 1988, Tanaka *et al.* 1988), giving evidence that MR (Fig. 2) may in some instances reveal lesions of brain tissue, which have not been demonstrated by CT, CT remains as the best investigation in the acute treatment of severe head injury.

CT – appearance: Contusions may appear on the CT as either hyperdense, hypodense or as mixed dense lesions: Figs. 1a to c. The size of these lesions ranges from barely visible alterations to lesions exceeding 5 cm in diameter (see page 206). The CT findings often vary with the posttraumatic interval (page 208). Early secondary lesions include hemispheric swelling and focal or generalized edema. These will be displayed on CT either as non-enhancing focal enlargements and/or hypodense areas (Clasen *et al.*

Fig. 1. a,b,c. Types of cerebral contusions on non-enhanced CT scans: a) Hypodense right frontal contusion one week after a fall on the occiput. (Fo., R.W., 27 y., Rö 3117/78); b) mixed-dense left fronto-temporal contusion five days after a fall when intoxicated with alcohol. (Schu., H., 56 y., Rö 2387/88); c) hyperdense right frontal contusion 16 hours after fall. (Neu, K.H., Rö 932/87)

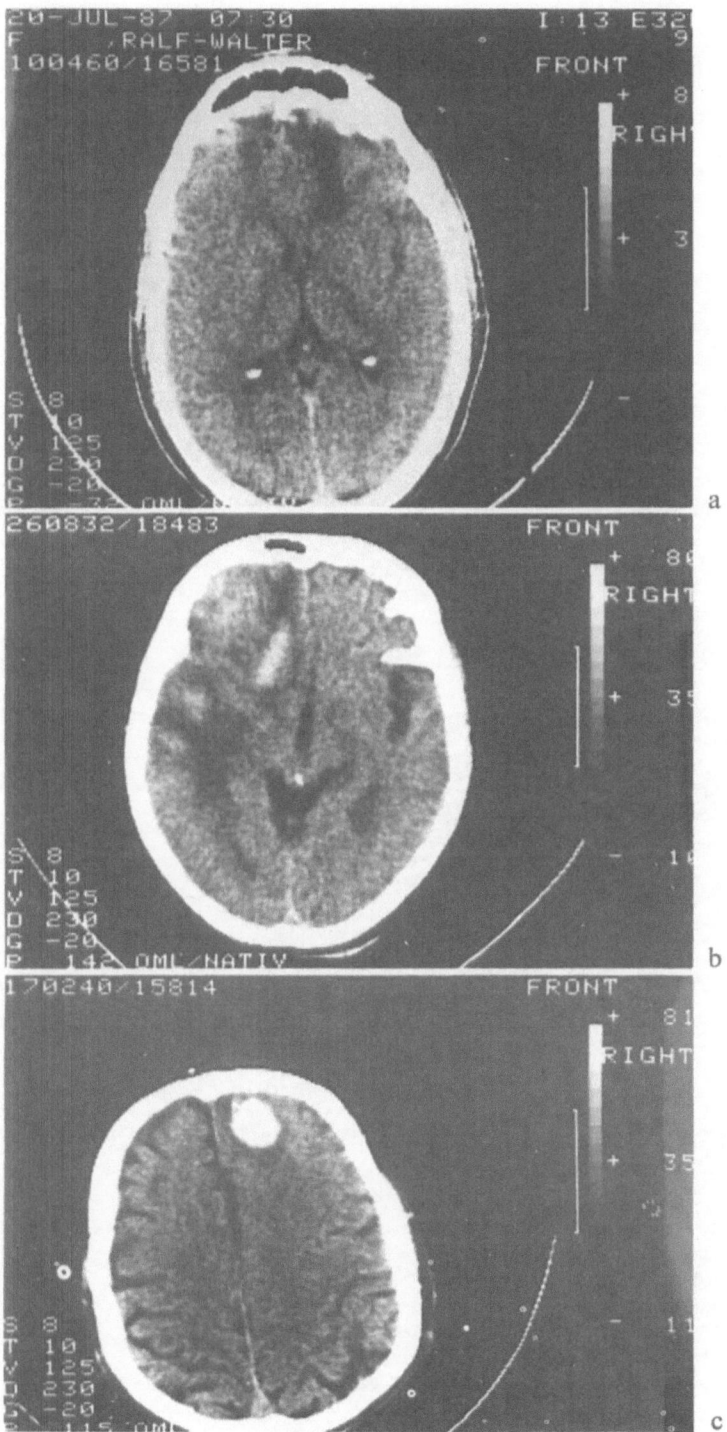

Fig. 1

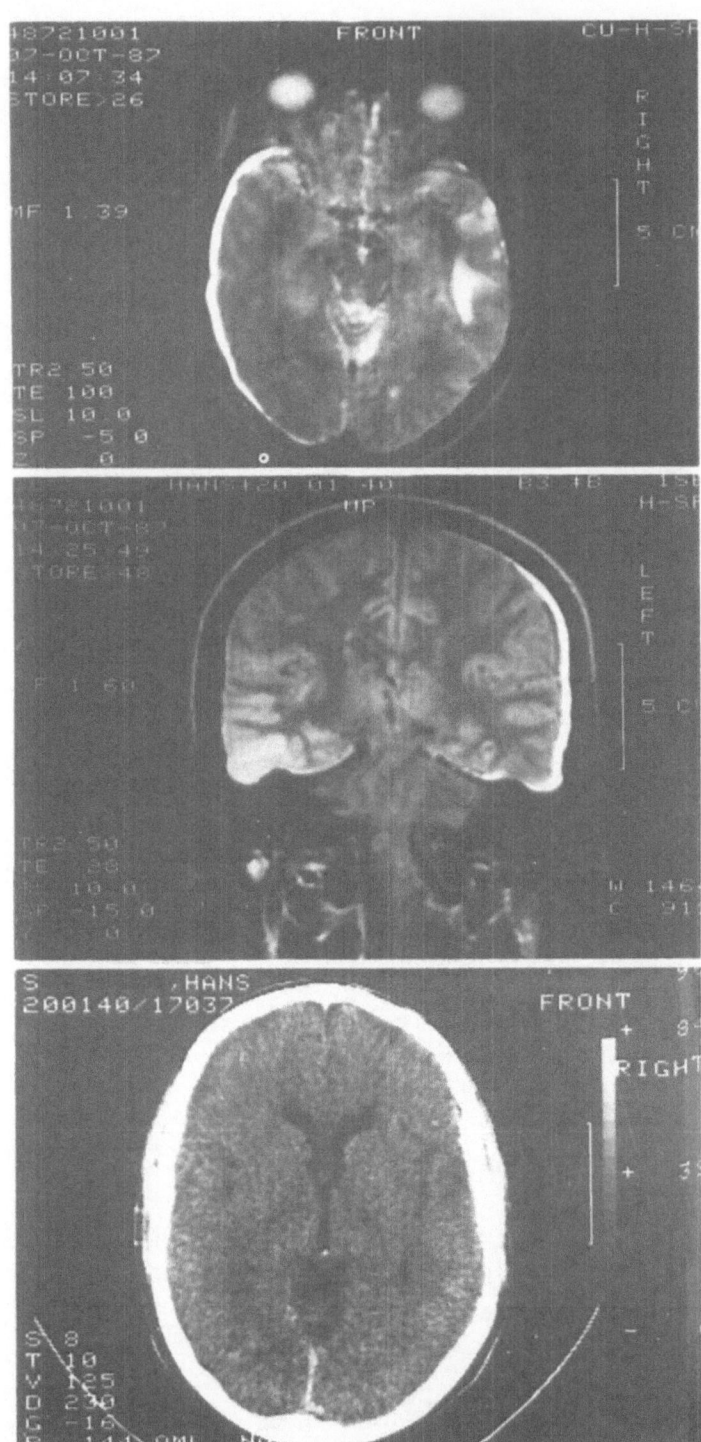

Fig. 2

1980, Penn and Clasen 1982). It will be shown on page 208 ff that about one third of the contusions may enlarge during the temporal course. But usually the initial high density and size of contusions gradually decrease. This resolution starts from the fifth day onwards (page 212). One to three weeks post-injury the proliferating blood vessels can produce contrast-enhancing contused areas. For a detailed explanation see Zimmermann and Bilaniuk 1984, Petersen and Voigt 1986, Kishore and Hall 1987, Bernasconi *et al.* 1988, and the chapters by Oprescu, Nakamura, Foroglou and Richard in this volume.

In this chapter only focal parenchymal contusions and lacerations are considered, excluding extracerebral hematomas and secondary lesions as mentioned above.

Correlation of Clinical Manifestation and CT-findings

Type of Lesions

As a rule, neurological deficits correspond with the size and location of a contusion as visualized by CT. There are, however, contusions, that are clinically silent, while there are also patients with severe neurological deficits without evidence of brain lesions on the CT scan, as cerebral dysfunction may occur in the absence of structural damage (Gennarelli *et al.* 1982, Miller and Becker 1982, Vigouroux and Guillermain in this volume).

In an attempt to quantify these phenomena, we analysed a series of *289 consecutive patients*, treated between 1979 and 1984, and we examined the correlation of the clinical manifestations with the CT findings suggestive of contusional hemorrhages: 141, *i.e.* 49% of the patients had brain contusions, lacerations and intracerebral hematomas.

Time of Investigation

91 patients were examined within 3 hours after injury, 24 further patients within 24 hours and 16 patients after one day.

In all but 11 patients, evidence of contusions was identified either on the initial or on a repeat CT scan: There were *120 contusions as main lesions* and 115 accompanying parenchymal alterations related to other extra or intracerebral lesions. In 10 patients an intracerebral hematoma had formed: Table 1.

Fig. 2. MR imaging of right temporo-basal cerebral contusion, clearly visible on CT scan. Small left parietal subdural effusion, 7 days after fall. (Schü., H., 47 y., Rö 4081/87)

Table 1. *Distribution of CT-findings in 289 Head Injuries*

Head injuries, CT investigated patients	289		
Permanent normal	11		
Main parenchymal lesions	130		
Contusions on first or repeat CT		120	
Secondary enlargement			33
Intracerebral hematomas		10	
Extracerebral hematomas	125		
Brain edema	16		
Intraventricular bleeding	7		
Accompanying lesions			115

Table 2. *Grading of Contusional Parenchymal Lesions*

Size of lesion on CT		
Group		No. of patients
A no contusion on initial CT scan: (25)		
A1: no contusion on repeat CT scan	11	
A2: new lesion on repeat CT scan		14
B less than 2 cm		38
C 2–5 cm		46
D more than 5 cm		22
		11 + 120 = 131

Size of Contusions

The *size* of the lesions was graded into 4 categories in Table 2 and Fig. 3:
Comparing the type and size of the lesions on the first, initial CT and on the repeat CT the larger hyperdense hemorrhagic contusions evidently become more frequent: Fig. 4.

Location of Contusions

The well known preferential location of contusions in the polar and basal frontal and temporal regions, due to the contact of the brain tissue with bony structures at the base of the skull, is already reported in the pathomorphologic findings by Oprescu in this volume. 12 out of the 120 cases had hemorrhagic deep midline lesions: Table 3 and Fig. 5.

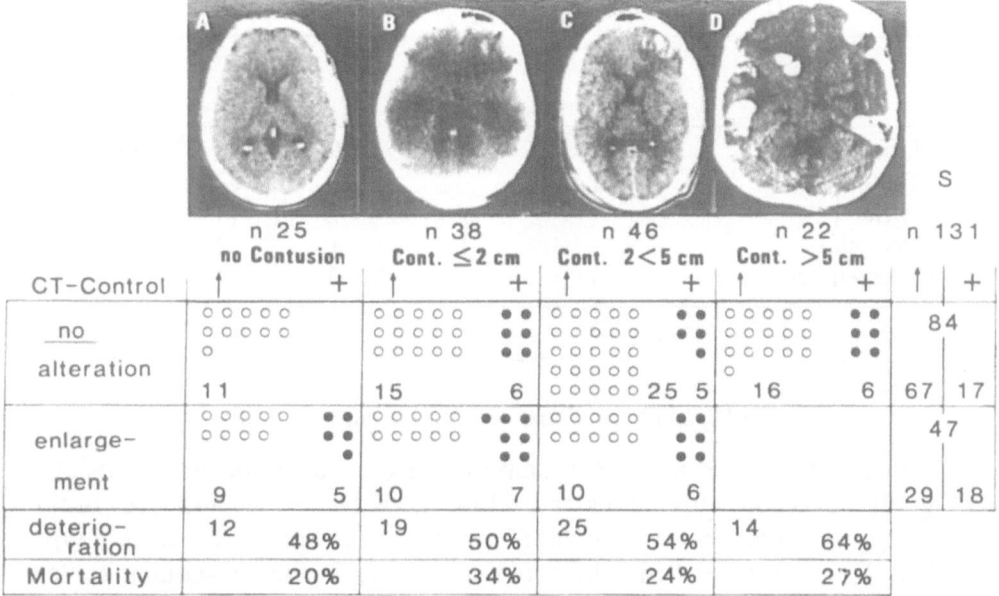

	n 25 no Contusion		n 38 Cont. ≤2 cm		n 46 Cont. 2<5 cm		n 22 Cont. >5 cm		n 131	
CT-Control	↑	+	↑	+	↑	+	↑	+	↑	+
no alteration	11		15	6	25	5	16	6	84 67	17
enlarge-ment	9	5	10	7	10	6			47 29	18
deterio-ration	12	48%	19	50%	25	54%	14	64%		
Mortality		20%		34%		24%		27%		

Fig. 3. Cerebral contusions of initial size, in four sub-groups, their frequency of enlargement and the mortality of 120 patients. A 2) Initial CT still normal, B) size below 2 cm, C) size 2 to 5 cm, D) size over 5 cm

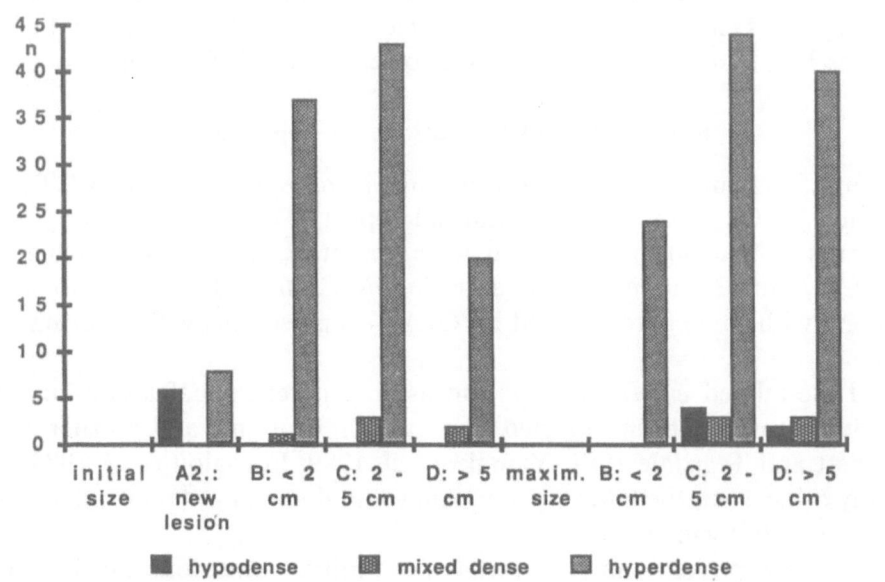

Fig. 4. Size and density of cerebral contusions on initial CT scans and their maximal appearance on repeat CT scans. A 2 = initial CT was normal, repeated CT showed new lesion. Meaning of B, C, D as in Fig. 3

Table 3. *Location of 120 Focal Contusions on CT*

	Left	Mid-line	Right	S	%
Frontal	18	11	19	48	40
Temporal	32	—	20	52	43
Parietal	11	—	3	14	12
Occipital	—	1	1	2	2
Infratentorial	1	—	3	4	3
	62	12	46	120	

Multiple Contusions, Accompanying Lesions

In 21 patients the initial CT demonstrated either multiple contusions (Fig. 6) or accompanying extracerebral hematomas, generalized brain swelling, or edema. Seven out of 17 small contusions were contre-coup-lesions. These cases will not be listed separately in the subsequent tables and figures.

State of Consciousness

The Glasgow Coma Scale Grading – GCS – (Teasdale and Jennett 1974) is used as compared to the Brussels Coma Scale – BCS –, which distinguishes clear consciousness from clouded consciousness, and 4 coma grades I to IV (Brihaye *et al.* 1978, Frowein *et al.* 1976, 1980).

Enlargement of Contusions on Repeat Scan

In 25 patients no conspicuous lesion was revealed on the first CT scan (group A). Of these, 11 had a normal repeat CT scan (group A1). Five patients with a normal CT scan were comatose: 2 were in coma grade I/GCS 7, and 3 patients in coma grade II/GCS 6–7. There was no single patient with coma grade III and IV/GCS 3–4 presenting with a normal CT scan.

Time-related *enlargement* of contusional parenchymal lesions has already extensively been reported by Lanksch, Grumme and Kazner 1979, Jennett and Teasdale 1981, Yamaki *et al.* 1989, Foroglou *et al.* 1989 and many others. But the intensity, frequency and dynamics of these alterations are less well documented.

In our experience, in two thirds of the initial contusions, *i.e.* 84 out of 131 patients without considering the time of the investigation, the follow-up CT scans remained unchanged or showed improvement; the mortality of these patients was 20% (Fig. 3).

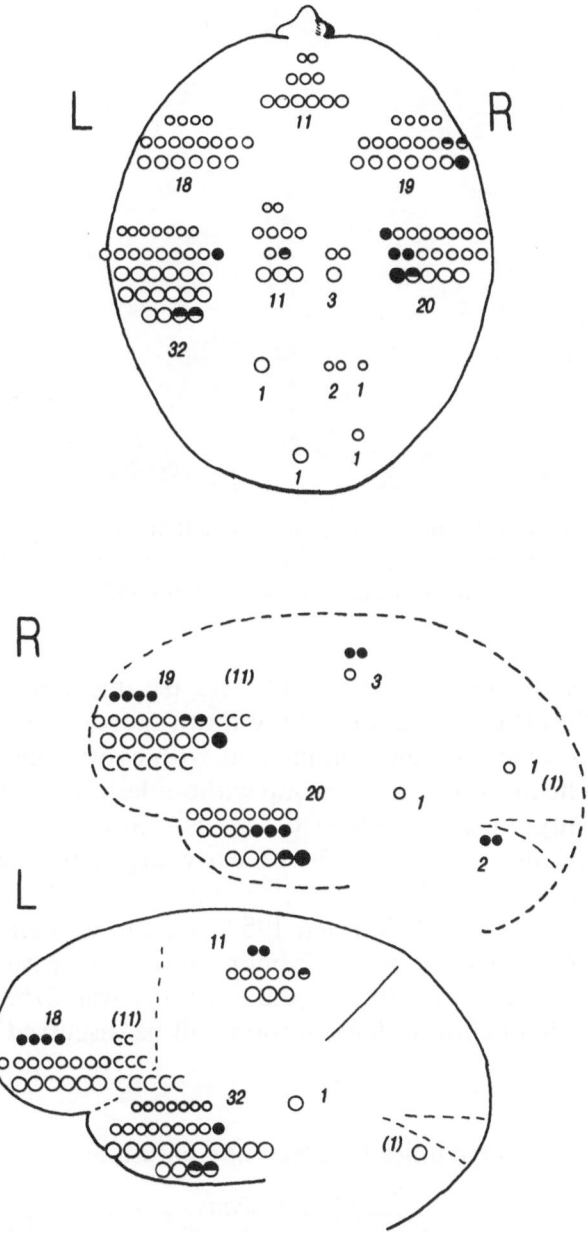

Fig. 5. The fronto- and temporo-polar regions were the preferred location of 120 contusions and lacerations on CT scan. 12 midline-lesions are represented by half-circles from the lateral view. Black dots: hypodense, semi-black dots: mixed-dense, fair dots: hyperdense lesions

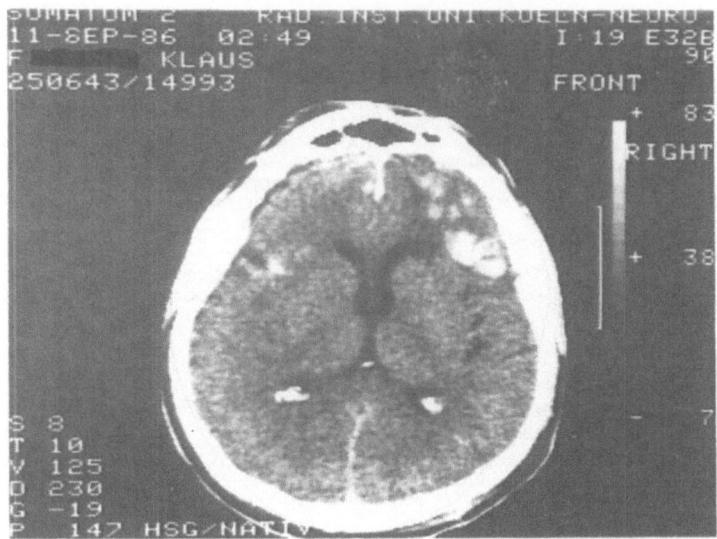

Fig. 6. Multiple cerebral contusions right and left fronto-temporal. Patient was found 2 hours earlier in a state of clouding of consciousness. Hemorrhagia out of the right ear. (Fie, K. 43 y., Rö 4062/86)

One third of all contusions, *i.e.* in 47 cases, became larger as seen on the repeat CT without regard to timing; the mean mortality of these patients was 38%, about twice as high as in patients with contusions not exhibiting enlargement. The mortality in the group without lesions in the initial but in the second or any other of the later CT scans – group A2 – was 20%, and thus about the same as in group D with early large contusions over 5 cm, which had a mortality of 27%.

Similarly, Espersen and Petersen 1982, relating outcome to CT main lesion, reported that in cases with no abnormalities the death rate was 16%, in patients with cerebral contusions the mortality was 27%.

The frequency of clinical deterioration will be discussed subsequently.

Dynamics of Cerebral Contusions

Early CT scans

Analysis of 80 early CT scans performed within 3 hours of the injury revealed a relation between size of lesion and the level of consciousness, as coma with neurological deficit was predominantly associated with lesions below 2 cm (group B, 13 patients), and with lesions 2 to 5 cm large (group C, 11 patients): Table 4, Fig. 7.

Table 4. *Distribution of Survivors and Fatal Courses of 80 Patients with Early CT Scans Performed Within 3 Hours of Injury: Relation Between Size of Lesion (A2 to D) and the Level of Consciousness* (see text page 210 and Figs. 10 to 13)

Consc.	GCS	BCS		A2	A2	A2	Ba	Bb	B	B	Ca	Cb	C	C	D	D	D	Total	Total	Total	Total
				Surv	Dead	Sum	Surv	Surv	Dead	Sum	Surv	Surv	Dead	Sum	Surv	Dead	Sum	Surv	Dead	Sum	Dead%
clear	13 to 15		Surv	1			1	2			4				1			9			
			Dead						1										1		
			Sum	1						4				4			1			10	10
clouded	9 to 12		Surv	2				2				1						5			
			Dead		2								1						3		
			Sum			4				2				2						8	38
Coma	7 to 8	Coma I	Surv	1			2	1			3	2			1			10			
			Dead		1				3				3						7		
			Sum			2				6				8			1			17	41
Coma	5 to 6	Coma II	Surv	1			7	3			4	3						18			
			Dead		1				3				4						8		
			Sum			2				13				11						26	31
Coma	4	Coma III	Surv	4				2			3	2						11			
			Dead		1				4							1			6		
			Sum			5				6				5			1			17	35
Coma	3	Coma IV	Surv																		
			Dead						1				1						2		
			Sum							1				1						2	100
Total			Surv	9			10	10			14	8			2			53			
			Dead		5				12				9			1			27		
			Sum			14				32				31			3			80	
			Dead%		36				38				31			33					34

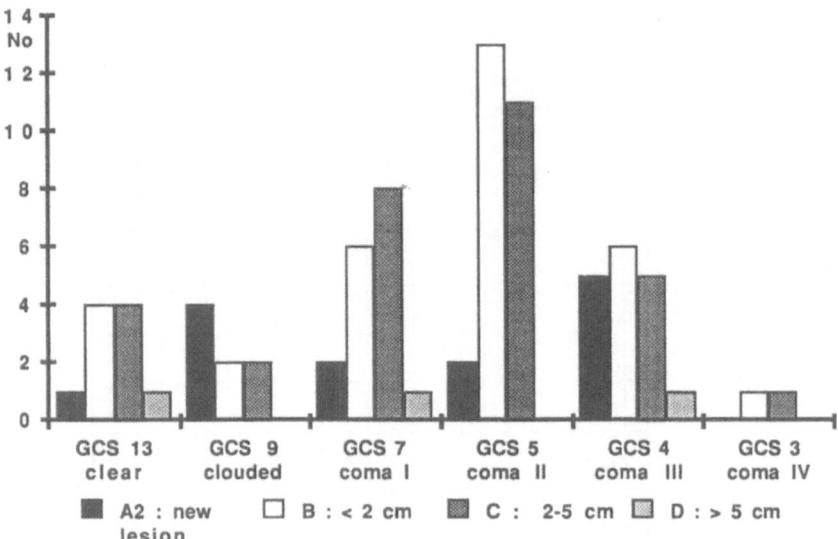

Fig. 7. Early contusions: Relation of size with the state of consciousness: 80 early
CT scans performed within 3 hours after injury

It is also evident from Fig. 8 that patients with normal CT scans in this
early period after injury, but subsequently with new lesions, A2, as well as
patients with initial lesions over 5 cm large, had about the same *mortality*
rate, between 33 and 36%. This fact is important for the timing of assessment (Fig. 9a and b).

Temporal Course

The clinical courses and CT findings differed in the sub-groups A, B,
C and D. Figs. 10–13:

Group A2: After an initially normal CT, the earliest lesion was revealed
8 hours after injury; the others were detected within 48, rarely after more
hours. Resolution of hyperdense lesions was seen from the 5th day on after
injury. In spite of the delayed appearance of parenchymal lesions, the
clinical syndrome improved finally in 9 out of 14 patients.

Group B: 19 out of 38 patients with lesions less than 2 cm large
deteriorated, in spite of unchanged size of the lesions. Enlargement of the
lesion was seen in 17 cases after 12 hours at the earliest. Resolution occurs
onwards from 5 to 9 days on. Mortality reached a maximum of 38%.

Group C: There were 46 lesions 2 to 5 cm large. Improvement of the
clinical syndrome was observed in 11 cases. Deterioration in 24 patients
occurred in spite of 30 unchanged medium-sized, mostly hyperdense

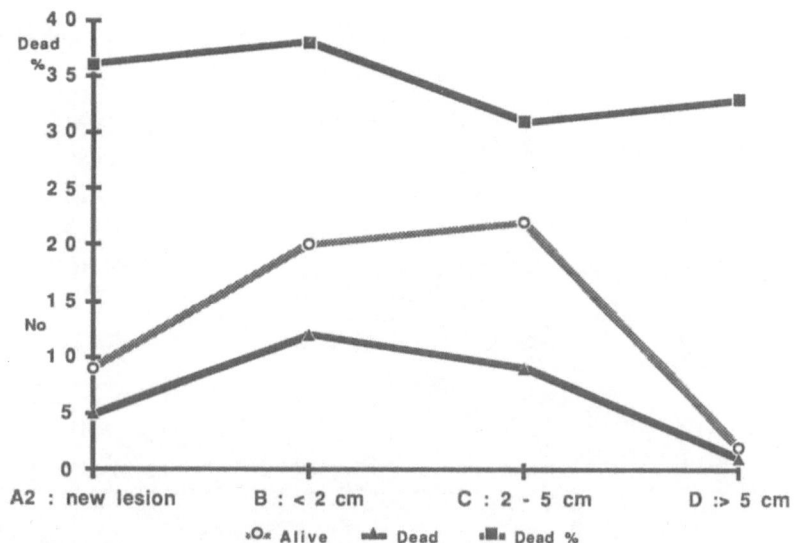

Fig. 8. Early contusions: 80 cases with CT performed within 3 hours after injury. In spite of the small numbers (*No*) the mortality does not vary very much between 31 and 38% with the size of lesions

lesions. 16 lesions were enlarged after 10 hours at the earliest and after 11 days at the latest. Resolution was seen from the 5th day up to the 21st day onwards.

Group D: 22 lesions measured more than 5 cm in size: In spite of 14 intercurrent deteriorations there were 16 survivors. In addition to the 3 patients investigated early (Table 4) 5 out of 6 fatal courses occurred with a severe coma grade II and III, corresponding to GCS 4 to 7. 16 lesions were enlarged at the earliest 11 hours later; 4 patients had operative treatment.

Altogether, 47 out of 120 contusions, *i.e.* 39%, either first evolved or were enlarged on the repeat CT scan at the earliest 8 hours later, in half of these cases between 12 and 36 hours later, and at the latest usually after 5 days, exceptionally after 11 to 12 days.

Intracerebral Hematoma

Frequency

The distinction of cerebral contusions forming a confluent hematoma from an intracerebral hematoma is not clear-cut and will be discussed in the chapter by Foroglou.

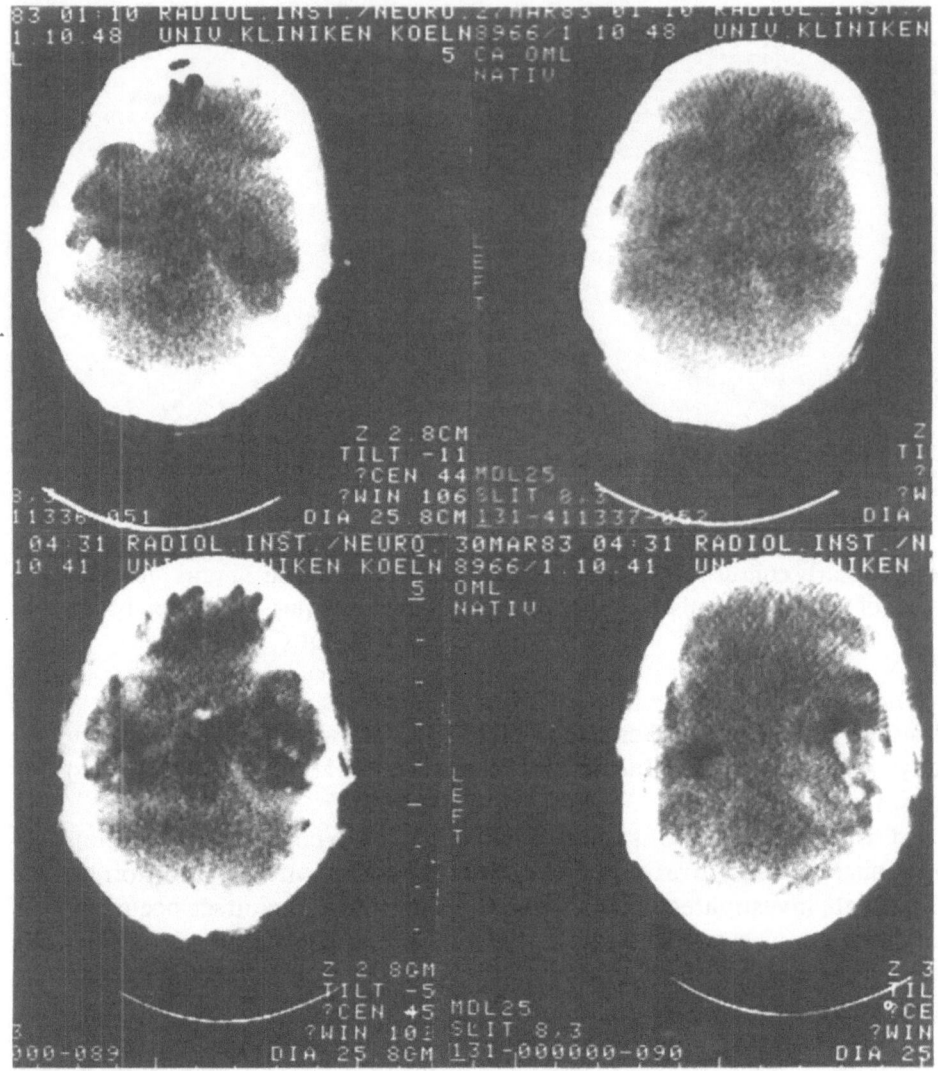

Fig. 9a. After initially normal CT scan 2 hours after a fight, the repeat CT scan on the third day revealed the hyperdense right temporal contusion (Gr., L.J., 35 y., Rö 1515/83)

Fig. 9b. Cerebral laceration and air-inclusion adjacent to a right temporal compound fracture, 2 hours after a compression-injury of the head. 8 hours later, a large contusional hemorrhage. Good outcome after surgery (E., Th., 38 y., K 1153/88)

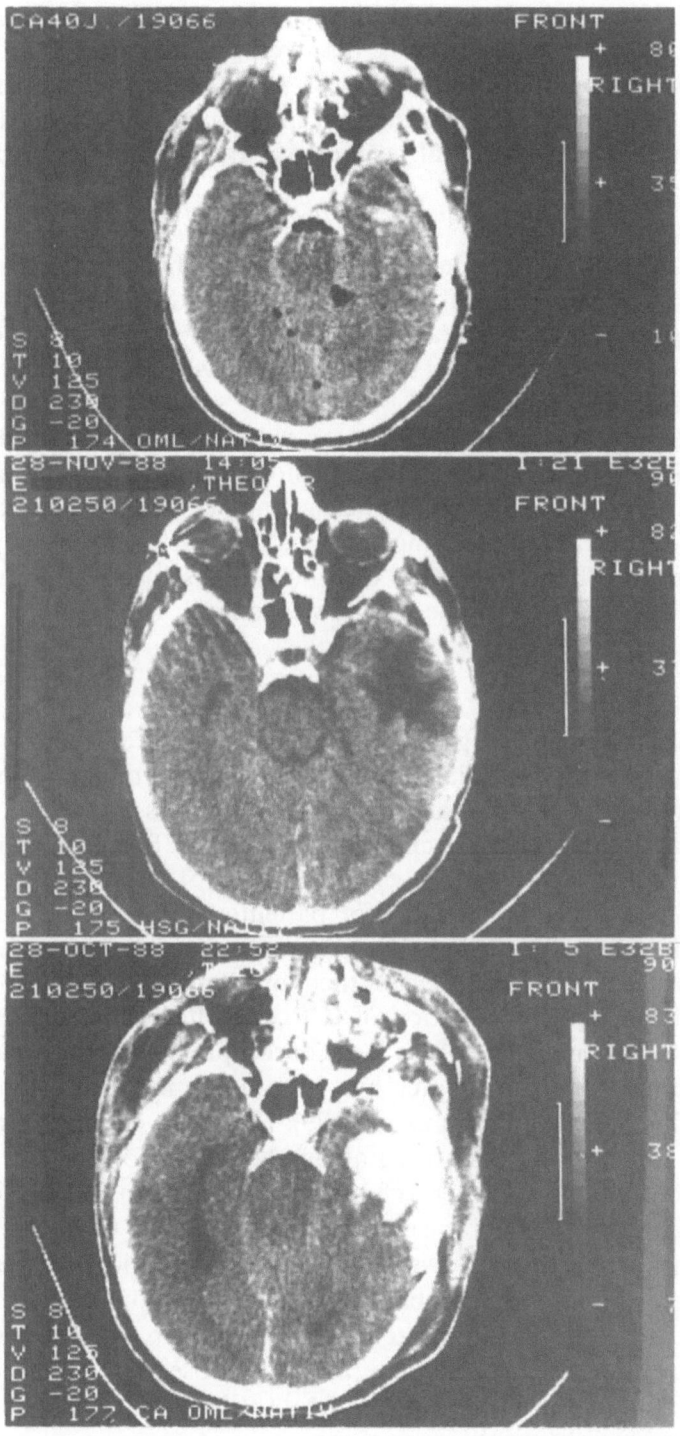

Fig. 9b

In only 10 out of 289 initial CT scans, about 3.5%, an intracerebral hematoma as a main lesion was discovered 1 to 5 days after injury. 9 patients were operated on; 3 comatose patients died after operation.

Only 4 out of all 120 contusions of immediate or delayed onset were forming confluent intracerebral hematomas, that required an operation.

Secondary Delayed Intracerebral Hematoma

In our opinion a true delayed intracerebral hematoma can be considered only if at least two CT scans were negative during the first 24 hours. This strong prerequisite is rarely fulfilled in the cases reported in the literature (Nanassis, Frowein *et al.* 1989, Zülch 1984; see also the chapter by Foroglou in this volume).

In one exceptional case of an 18-year-old patient treated at our department (1634–84) the absence of conspicuous findings was confirmed by repeat CTs 2, 8 and 24 hours after head injury. Sudden deterioration of consciousness and the appearance of a right sided hemiparesis prompted a third repeat CT on the 12th day, which disclosed an intracerebral hematoma, revolving spontaneously 2 weeks later: Fig. 14.

Treatment

Intensive treatment is guided by the neurological syndrome, EEG, evoked potentials, transcranial Doppler measurement of bloodflow velocity and ICP-monitoring, as described in the chapters by Vigouroux and Guillermain, Richard and Firsching in this volume. See also Marshall and Marshall 1987.

Operation is urgently required for patients with large contusions and considerable mass effect; they should have a craniotomy at the time of presentation to minimize the development of edema and herniation. But the duration of subsequent reduction of ICP was often short in our patients and in that of others. Therefore, a large decompressive craniotomy with or without opening the dura-mater has been recommended (Heppner and Argyropoulos 1972, Yamaura *et al.* 1979). However, compared with more conventional resection of contused areas, there was no clear-cut improvement in the outcome (Cooper 1987, Karlen and Stula 1987). Sisco *et al.* 1988, conclude, that hemicraniectomy has little to add to the treatment of unilateral hemispheric brain swelling in a general application, but it could be considered in rare cases. Lobato 1988, shares the opinion that must be considered in severely head-injured patients who do not show brainstem dysfunction immediately after injury, and who subsequently develop neurological deterioration with otherwise uncontrollable intracranial hypertension from cerebral hemispheric swelling.

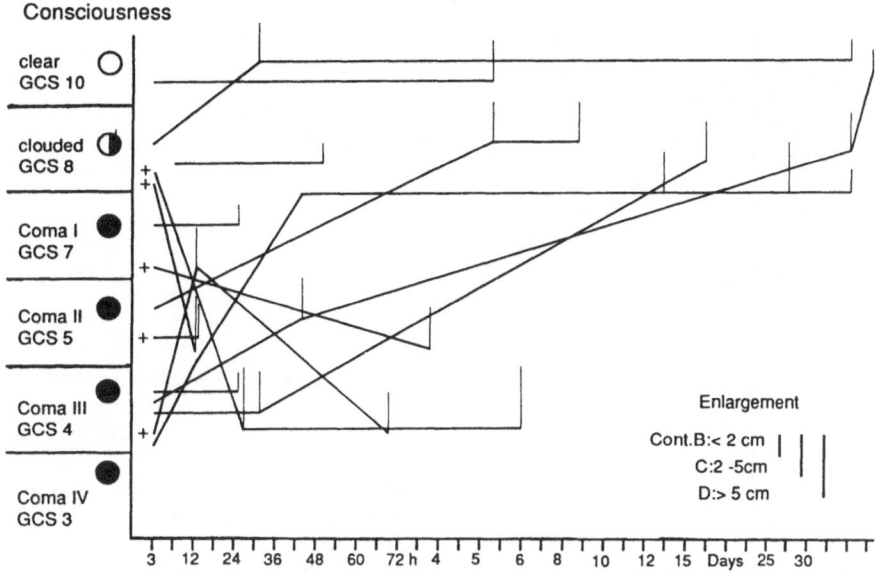

Fig. 10. Time-course of clinical syndrome and size of CT-revealed contusions in the sub-group A2: initial CT scan was normal, subsequent appearance of parenchymal lesions of variable size

Prognosis

Mortality

The mortality of comatose patients with *early* CT scans did not differ very much with the size of the lesions: 31 to 41% in the individual coma grades, as shown in Table 4 and Fig. 8.

Lobato *et al.* 1988, reported an overall mortality of single contusions of 14% (3 out of 14 cases) and 54% (36 out of 62) in multiple contusions, while diffuse brain injury had a mortality of 28% (27 out of 96 cases). Marshall *et al.* 1977, found a midline shift of more than 15 mm a reliable prognosticator of poor outcome.

It is evident from a study of all our patients with all possible size of contusions at different time periods (Fig. 15), that an assessment after the first 3 hours for patients with clouding of consciousness and those with GCS 7/coma grade I, shows an unusually high mortality of 18 and 40% respectively. After 6 hours the mortality of these two groups diminishes to the usual low mortality of less than 10%, but it increases in patients with the more severe coma grades II and III, GCS less than 6, to over 50%. This is in accordance with the report by Vigouroux and Guillermain in this volume, page 91.

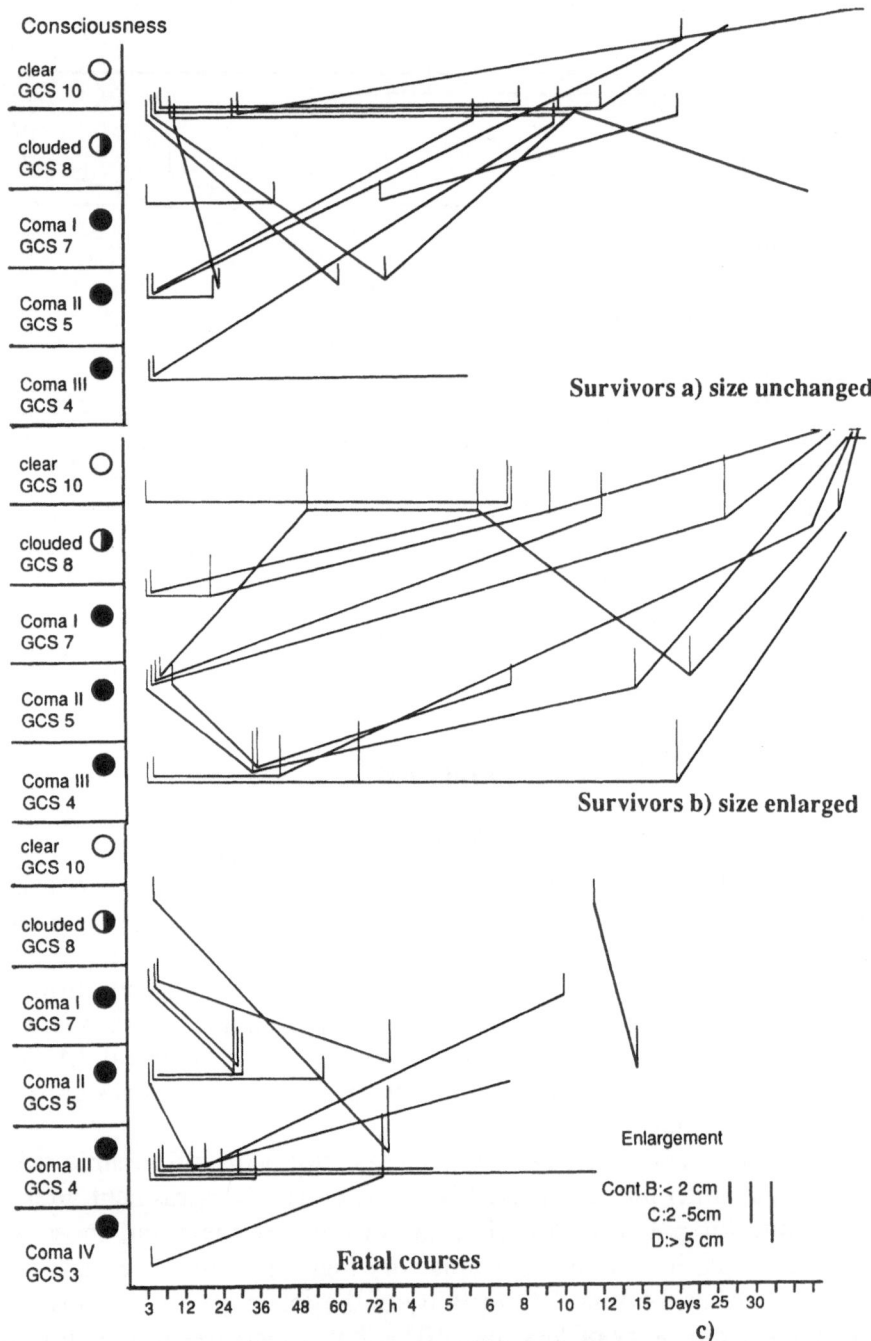

Fig. 11a, b, c. Temporal courses in patients with initial lesions less than 2 cm large
(group B); a) survivors who did not have an enlarging lesion on repeat CT scan,
b) survivors with an enlarging contusion, c) fatal courses

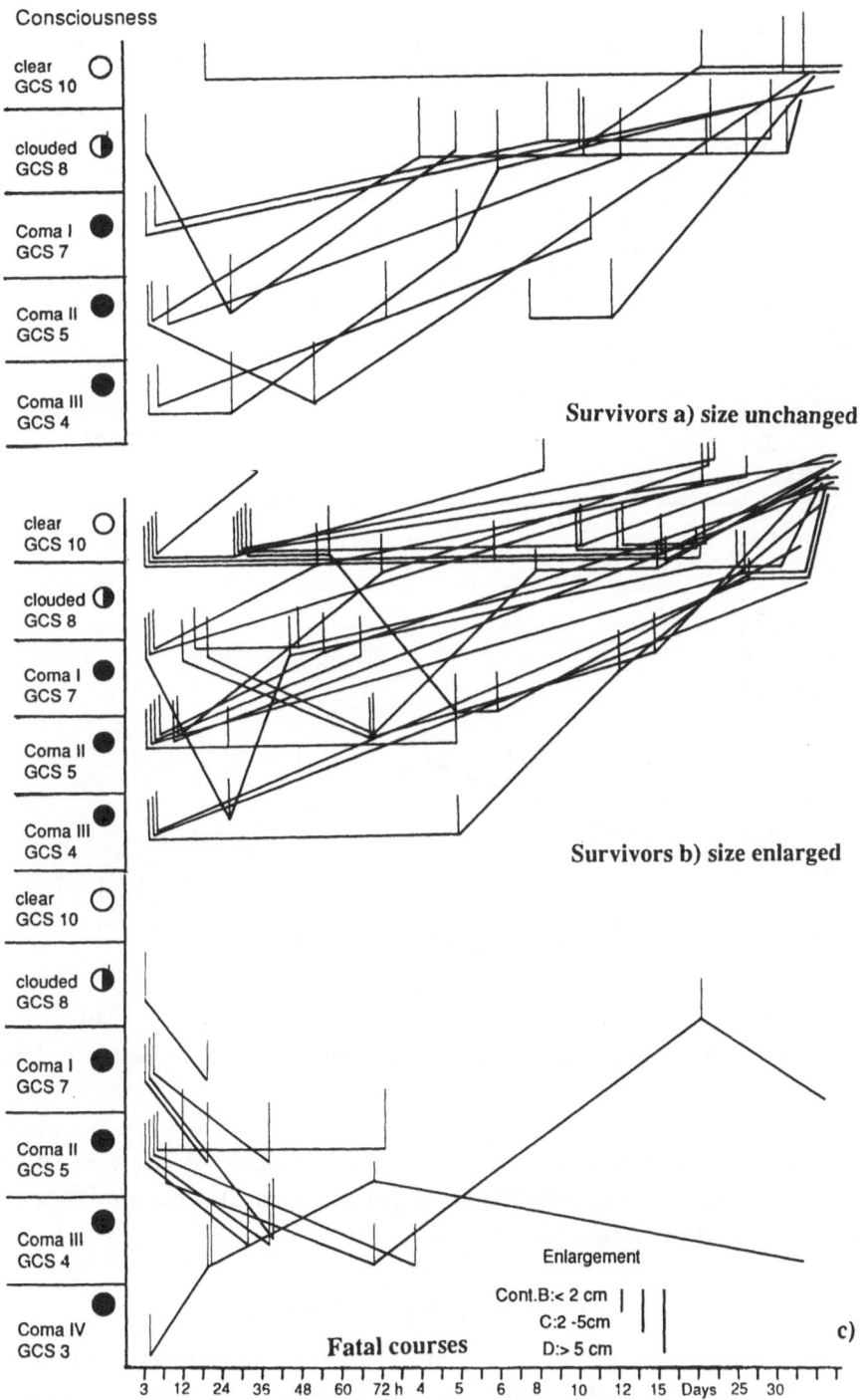

Fig. 12a, b, c. Temporal courses in patients with initial lesions 2 to 5 cm large (group C): a, b, c as in Fig. 11

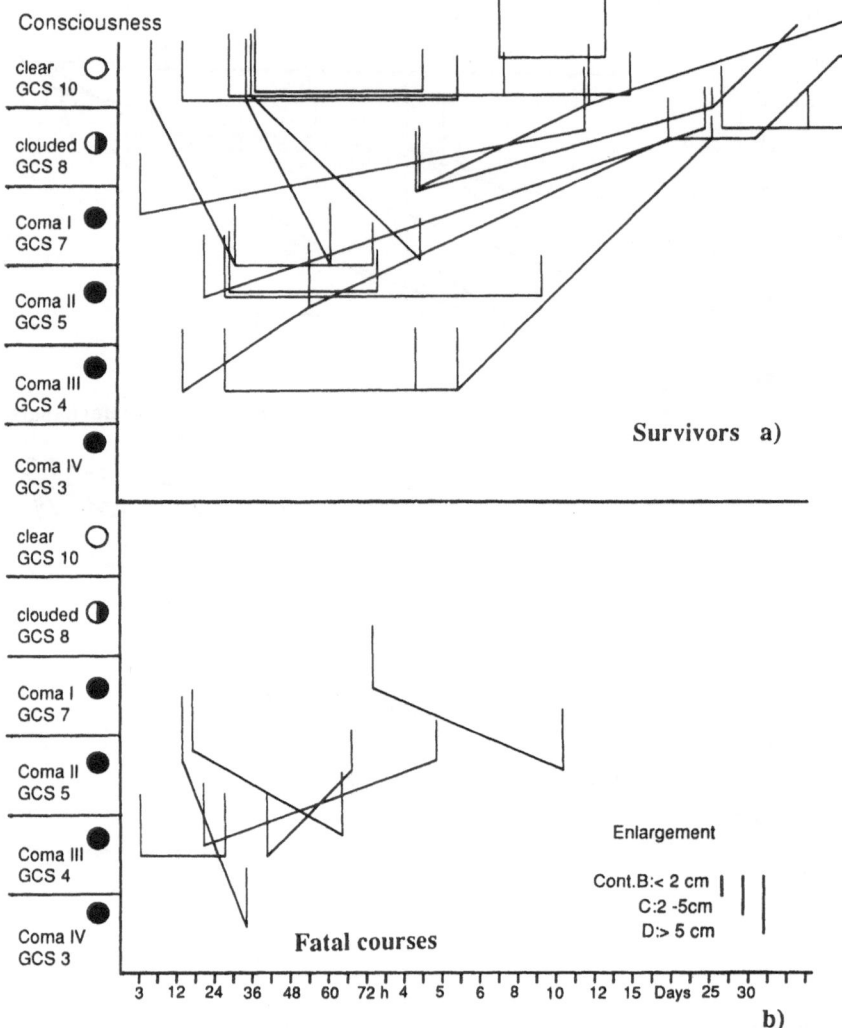

Fig. 13a and b. Temporal courses in patients with initial lesions larger than 5 cm
(group D)

The predictive value of the *level of consciousness* of any size of contusions was therefore concluded to depend on the timing and becomes
realistic not earlier than 24 hours after injury.

Time-related Manifestations

Based on the timing of the initial CT, 409 cases with either epidural,
subdural or intracerebral hematomas or enlarged contusions were analysed
(Fig. 16). There were only 3 large contusions on the first CT. In the first

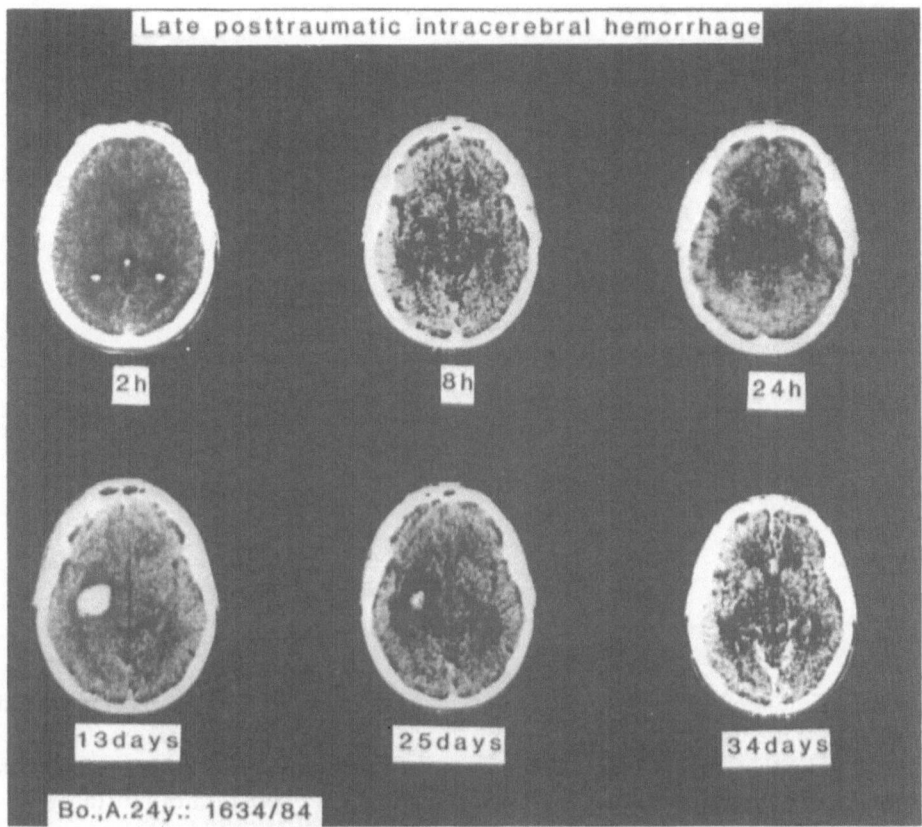

Fig. 14. Delayed intracerebral hematoma: initial and two repeat CT scans negative. After 12 days acute clinical deterioration and the appearance of a deep situated small hematoma. Conservative treatment. Recovery (K 1634–84)

three hours mostly acute subdural and, to a lesser degree, epidural hematomas predominate, decreasing rapidly in number with the interval after the injury. From the 6th to 12th hour onwards the enlarged contusions are more frequent, continuing in the later part of the first and then on the second day.

Time-related Probability

The *time-related probability* of encountering any of these focal space occupying lesions is derived from the total number of patients with CT investigations at different time periods after the injury (Fig. 17): The acute hematomas make up the peak-probability of 50% in the second hour after injury. The small peak 12 hours after injury is due to delayed diagnosis of

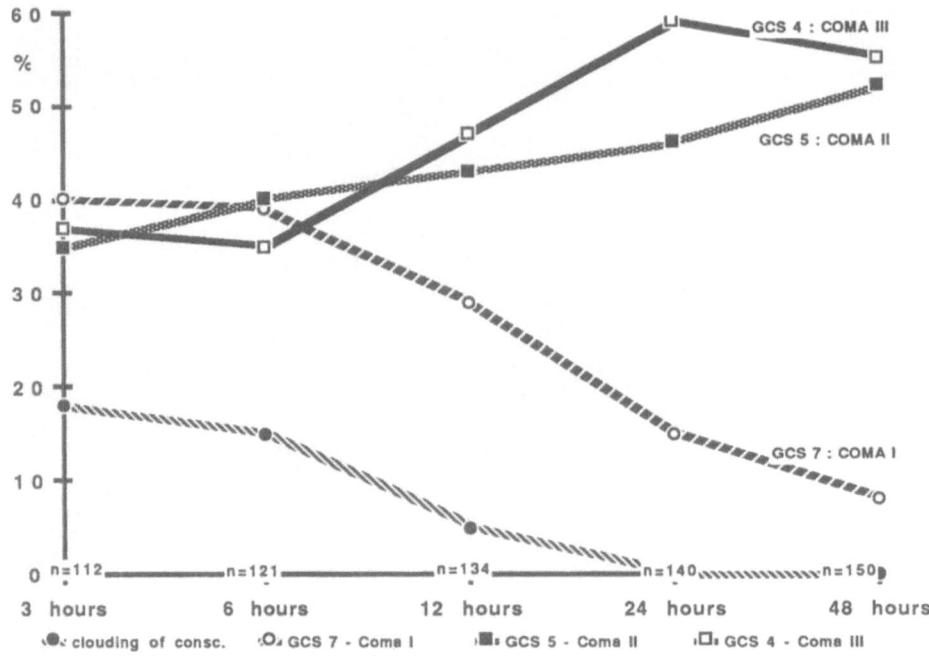

Fig. 15. Time-related assessment of contusional syndrome: Mortality calculated from 155 patients with different coma grades at different posttraumatic time-intervals: In the earliest stage, 3 hours after injury, in patients presenting with clouding of consciousness and with coma grade I the prediction of mortality was too high, in patients with coma grades II and III it was too low

epidural and subdural hematomas and also from the early enlargement of contusions.

The risk of an enlargement of contusions continues up to 48 hours, in decreasing probability up to 12 days.

The resulting overall mortality, as related to the timing of CT, reaches a maximum within the first 3 hours after injury, and there is a second sharp increase 6 to 12 hours after injury for focal space occupying lesions (Fig. 18).

Summary: How to Do It

In a series of 289 patients admitted to our departments after a head injury between 1979 and 1984, computerized tomography (CT) revealed cerebral contusions and lacerations in 49% (141 cases) (Table 1). Gennarelli *et al.* 1982, observed 205 focal lesions out of 1,107 severely injured patients in seven centers, that is a frequency of 19%. In 14 out of our 120 patients with repeat-CT scans, the contusion had been missing on the initial CT.

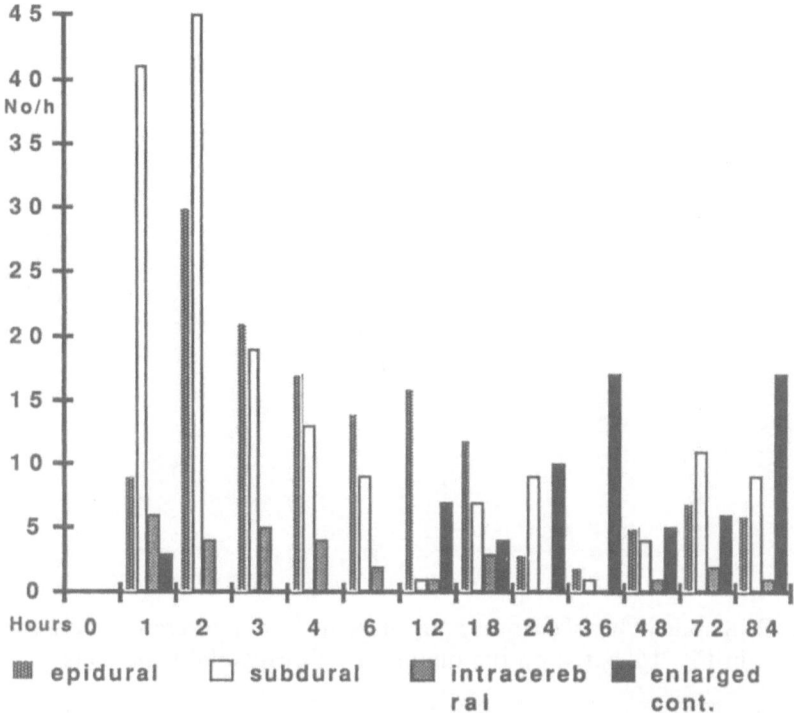

Fig. 16. Time-related manifestation of 409 posttraumatic acute focal intracranial mass-lesions: total number of intracranial hematomas and cerebral contusions either initially present or enlarged as seen on CT scans at various intervals after injury

One-third of all cases showed enlargement of the hemorrhagic lesions on the repeat-CT scan.

This dynamic evolution of contusions proved to be unsystematic, unpredictable and not clearly related to the clinical syndrome.

It is obvious that any rapid deterioration of the neurological syndrome, especially the level of consciousness, should give rise to an immediate repeat-CT, no matter how little time has elapsed since the most recent CT.

The interval, however, after which a CT scan is to be repeated in clinically unchanged patients, is controversial. Yoshimizu *et al.* 1989, advise an interval of two hours after the initial CT. As we have found significant enlargement of contusions from 8 hours onwards, we recommend a repeat-CT scan after 8 hours as a rule. The risk of enlargement of the contusions continues up to 48 hours and exceptionally up to 12 days.

Ten patients were found to have an intracerebral hematoma on the initial CT scan, out of 289 patients, *i.e.* 3.5%. Four out of 120 patients, *i.e.*

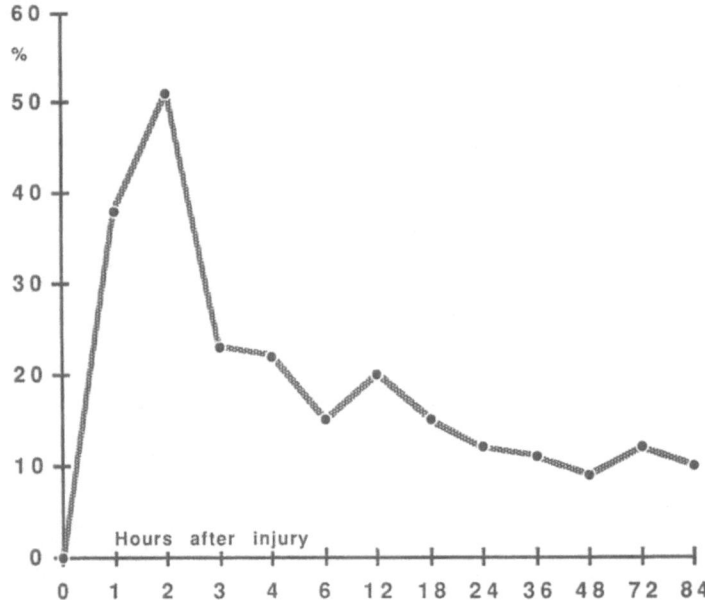

Fig. 17. Probability (%) of encountering an acute focal intracranial mass-lesion, as in Fig. 16, based on the timing of CT scans of 409 patients

3%, exhibited an intracerebral hematoma from previously contused or lacerated tissue.

Including the other neurosurgically relevant focal space occupying lesions, *i.e.* acute epidural and subdural hematomas, the probability of encountering such hematomas, large or enlarged contusions in posttraumatic comatose patients reached a maximum of 51% one to three hours after injury, and reached a second peak of 20% eight to twelve hours after the injury.

This temporal course is associated with similar but more prominent peaks of mortality, if any of these lesions were identified: 76% two to three hours after injury, and 60% at about 12 hours after trauma (Fig. 18).

Awareness of these important periods of highest risk following injury, should guide the surgeon to an appropriate posttraumatic regimen of neurological and radiological follow up investigations after quite short intervals.

This provides a concise and comprehensive interpretation and definition of the main types and time-relationships of the dynamic evolution of posttraumatic contusions and intracerebral hematomas (Fig. 19): the contusions decreasing in size, remaining unchanged for some days, or quickly enlarging; the intracerebral hematomas being rarely immediate, primary lesions, but usually developing after 8–24 hours, and only exceptionally being truly delayed for 4 or more days.

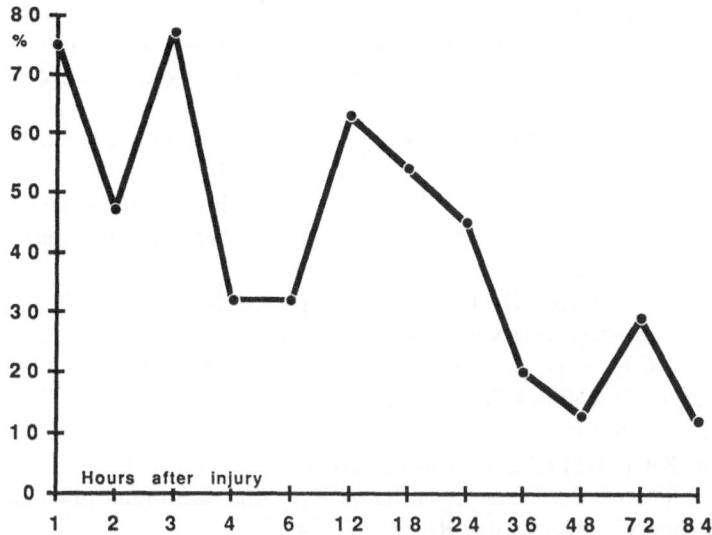

Fig. 18. Overall mortality (%) of acute focal intracranial mass-lesions, as in Figs. 16 and 17, related to a total of 780 head injured comatose patients at risk.

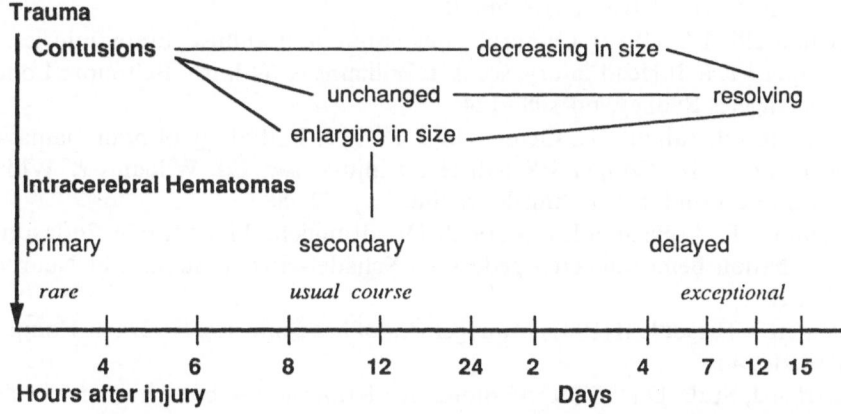

Fig. 19. Posttraumatic cerebral contusions may, with time, decrease in size, remain unchanged for several days, or they may quickly enlarge. Traumatic intracerebral hematomas are rarely found immediately as primary lesions, they usually develop after 8 to 24 hours and only exceptionally are they visualized on a repeat CT scan after four or more days

References

1. Bernasconi V, Farabola M, Vaccari U, Bettinelli A, Sina C, Tomei G, Sganzerla E, Spagnoli D, Guerra P (1988) Neuroradiology of posttraumatic diffuse cerebral lesions. In: Montorsi M, Granelli P (eds) Surgical updating, Lecture Book III. Monduzzi Editore, Milano, pp 1520–1522

2. Brihaye J, Frowein RA, Lindgren S, Loew F, Stroobandt G (1978) Report on the meeting of the WFNS Neurotraumatology Committee, Brussels, 19th–23rd of Sept. 1976. I. Coma Scaling. Acta Neurochir (Wien) 40: 181–186

3. Clasen R, Huchman M, von Roenn K, Pandolfi S, Laing I, Clasen J (1980) Time course of cerebral swelling. Neurol 28: 395–412

4. Cooper PR (1982, 2nd ed 1987) Posttraumatic intracranial mass lesions. In: Cooper PR (ed) Head injury. Williams & Wilkins, Baltimore London Los Angeles Sydney

5. Espersen JO, Petersen OF (1982) Computerized Tomography (CT) in patients with head injuries. Assessment of outcome based upon initial clinical findings and initial CT scans. Acta Neurochir (Wien) 65: 81–91

6. Foroglou G, Patsalas I, Kontopoulos B (1989) The timing of CT. Neurosurg Rev 12 [Suppl 1]: 169–174

7. Frowein RA (1976) Classification of coma. Acta Neurochir (Wien) 34: 5–10

8. Frowein RA, auf der Haar K, Terhaag D (1980) Assessment of coma. Reliability of prognosis. Neurosurg Rev 3: 67–74

9. Gennarelli TA, Spielman G, Langfitt T, Gildenberg P, Harrington T, Jane J, Marshall L, Miller J, Pitts L (1982) Influence of the type of intracranial lesion on outcome from severe head injury. A multicenter study using a new classification system. J Neurosurg 56: 26–32

10. Gennarelli TA (1987) Cerebral concussion and diffuse brain injuries. In: Cooper PR (ed) Head injury, sec. Ed. Williams & Wilkins, Baltimore London Los Angeles Sydney, pp 108–124

11. Graham DI, Adams JH, Gennarelli TA (1987) Pathology of brain damage in head injury. In: Cooper PR (ed) Head injury, sec. Ed. Williams & Wilkins, Baltimore London Los Angeles Sydney, pp 72–88

12. Heppner F, Argyropoulos G (1972) Die dringliche bitemporale Entlastungstrepanation beim schweren gedeckten Schädel-Hirn-Trauma. Zbl Neurochir 33: 210

13. Jennett I, Teasdale B (1981) Management of head injuries. FA Davis Company, Philadelphia

14. Karlen J, Stula D (1987) Dekompressive Kraniotomie bei schwerem Schädelhirntrauma nach erfolgloser Behandlung mit Barbituraten. Neurochirurgia 30: 35–39

15. Kishore PRS, Hall JA (1987) Radiographic evaluation. In: Cooper PR (ed) Head injury, sec. Ed. Williams & Wilkins, Baltimore London Los Angeles Sydney, pp 51–71

16. Langfitt T, Obrist W, Alavi A, Grossman R, Zimmerman R, Jaggi J, Uzzeli B, Reivich M, Patton D (1986) Computerized tomography, magnetic resonance imaging, and positron emission tomography in the study of brain trauma. J Neurosurg 643: 760–767

17. Lanksch W, Grumme T, Kazner E (1979) Computed tomography in head injuries. Springer, Berlin Heidelberg New York

18. Levin HS, Williams D, Crofford MJ, High WM, Eisenberg HM, Amparo EG,

Guinto FC, Kalisky Z, Handel St F, Goldman AM (1988) Relationship of depth of brain lesions to consciousness and outcome after closed head injury. J Neurosurg 69: 861–866

19. Lobato RD (1988) Hemicraniectomy in the management of posttraumatic brain swelling. J Neurosurg 69: 963
20. Lobato R, Rivas J, Cordobes F, Alted E, Perez C, Sarabia R, Cabrera A, Diez I, Gomez P, Lamas E (1988) Acute epidural hematoma: An analysis of factors influencing the outcome of patients undergoing surgery in coma. J Neurosurg 68: 48–57
21. Marshall L, Smith R, Shapiro (1977) The outcome with aggressive treatment in severe head injuries. I. The significance of intracranial pressure monitoring. J Neurosurg 50: 20–25
22. Marshall LF, Marshall SB (1987) Medical management of intracranial pressure. In: Cooper PR (ed) Head injury, sec. Ed. Williams & Wilkins, Baltimore London Los Angeles Sydney, pp 51–71
23. Miller J, Becker D (1982) General principles and pathophysiology of head injury. In: Youmans JR (ed) Neurological surgery. Saunders, Philadelphia London Toronto Mexico City Rio de Janeiro Sydney Tokyo
24. Nanassis K, Frowein RA, Karimi A, Thun F (1989) Delayed posttraumatic intracerebral bleeding. Delayed posttraumatic apoplexy: "Spätapoplexie". Neurosurg Rev 12 [Suppl 1]: 243–251
25. Ommaya A, Rockoff DS, Baldwin M, Friauf WS (1964) Experimental concussion. J Neurosurg 21: 249–265
26. Ommaya A (1982) Mechanisms of cerebral concussion, contusions and other effects of head injury. In: Youmans JR (ed) Neurological surgery, Vol IV. Saunders, Philadelphia London Toronto Mexico City Rio de Janeiro Sydney Tokyo, pp 2877–1895
27. Penn R, Clasen R (1982) Traumatic brain swelling and edema. In: Cooper P (ed 9) Head injury. Williams & Wilkins, Baltimore London, pp 233–255
28. Petersen D, Voigt K (1986) Traumatische kraniozerebrale Erkrankungen. In: Schinz (ed) Radiologische Diagnostik, Bd. V, Teil 1. Thieme, Stuttgart New York, pp 572–594
29. Rowbotham GF (1942, 4th ed. 1964) Acute injuries of the head. Livingstone, Edinburgh
30. Sisco AG, Tippett II TM, Chapleau ChE (1988) Management of posttraumatic brain swelling. J Neurosurg 69: 962–963
31. Tanaka T, Sakai T, Uemura K, Teramura A, Fujishima I, Yamamoto T (1988) MR imaging as predictor of delayed posttraumatic cerebral hemorrhage. J Neurosurg 69: 203–209
32. Teasdale G, Jennett B (1974) Assessment of coma and impaired consciousness. Lancet II: 81–84
33. Yamaki T, Ueguchi T, Kuboyama T, Higuchi T, Hirakawa K (1989) Chronological evaluation of the occurrence and growth of the traumatic intracerebral hematoma. Neurosurg Rev 12 [Suppl 1]: 149–152
34. Yamaura A, Uemura K, Makino H (1979) Large decompressive craniectomy in

management of severe cerebral contusion. A review of 207 cases. Neurol Med Chir (Tokyo) 19: 717–728

35. Yoshimizu N, Hiramoto M, Notani M (1989) Study of those cases which showed rapid deterioration within a few hours after head injury – importance of follow-up CT scans at an early stage. Neurosurg Rev 12 [Suppl 1]: 175–177

36. Zimmerman RA, Bilaniuk LT (1984) Head trauma. In: Rosenberg RN (ed) The clinical neurosciences IV. Churchill Livingstone, New York Edinburgh London Melbourne, pp 501–508

37. Zülch KJ (1984) Die traumatische Spätapoplexie. Fortschr Neurol Psychiat 53: 1–24

Evoked Potentials in Head Injury

R. Firsching

Neurochirurgische Universitäts-Klinik, Köln
(Federal Republic of Germany)

With 3 Figures

Contents

Introduction

Monitoring evoked potentials offers the unique opportunity to obtain objective data on the function of various sensory neural pathways in comatose patients. The term 'evoked potentials' denotes bioelectrical responses caused by a stimulus. 'Evoked responses' are used synonymously with 'evoked potentials'. Three various kinds of evoked potentials are currently in use: somatosensory evoked responses (SER) after stimulation of a peripheral nerve, visual evoked responses (VER) after optic stimulation and auditory evoked responses (AER) after acoustic stimulation. From the alteration, loss or preservation of these evoked potentials one can draw some conclusions concerning the integrity of somatosensory, visual and auditory pathways. In head injury this information is of particular interest for the early assessment of the prognosis and for the diagnosis of brain death.

Anatomical and Neurophysiological Considerations

Somatosensory Evoked Potentials (SER)

Cortical somatosensory evoked responses are particularly prominent after electrical stimulation of the median nerve (Stöhr et al. 1982). Therefore median nerve stimulation has been most commonly used in SER studies, unless other peripheral nerves are of special interest. After stimulation, the arrival of the volley may be monitored at various parts of the somatosensory pathway. Electrodes placed at the neck (segment C_2) normally show a first negative deflection about 14 ms (N14) after the stimulus (Fig. 1a). Electrodes placed over the contralateral scalp will have a first negative deflection approximately 20 ms (N20) after the stimulus (Halliday 1982). As the generator of N14 is attributed to the upper cervical cord and the generator of N20 is believed to be structures near the cortex, the interval between N14 and N20 is called central conduction time (CCT) (Cant et al. 1986, Hume et al. 1979). The generator of later components of the median nerve SER remains uncertain. SER are divided into short (analysis time after stimulus up to 30 ms), middle (analysis time between 30 and 70 ms) and long latency (analysis time longer than 70 ms) SER. The normal latency of N20 and the normal range of the CCT has been used to distinguish normal from pathological short latency SER. Intervals longer than normal range indicate functional lesions of the somatosensory pathway. If the N14 component were delayed, while CCT was normal, a peripheral lesion would be likely. For a cortical SER, the volley has to pass through the entire length of the spinal cord, brain stem, thalamus on the way to the cortex. By contrast, the auditory pathway only passes through part of the brain stem and the thalamus on the way to the cortex, while the visual pathway does not pass through the brain stem at all.

Visual Evoked Response (VER)

Various kinds of optic stimulation are in use. Flash stimulation with light emitting diodes mounted in goggles is probably most practical in comatose patients. The flash must be powerful enough to penetrate the closed eyelids. The initial electrical response of the retina occurs after approximately 20 to 40 ms. Later components of the VER (Fig. 1b), as it is recorded from occipital scalp electrodes, are attributed to intracranial parts of the visual pathway. It remains unclear, however, which of the later

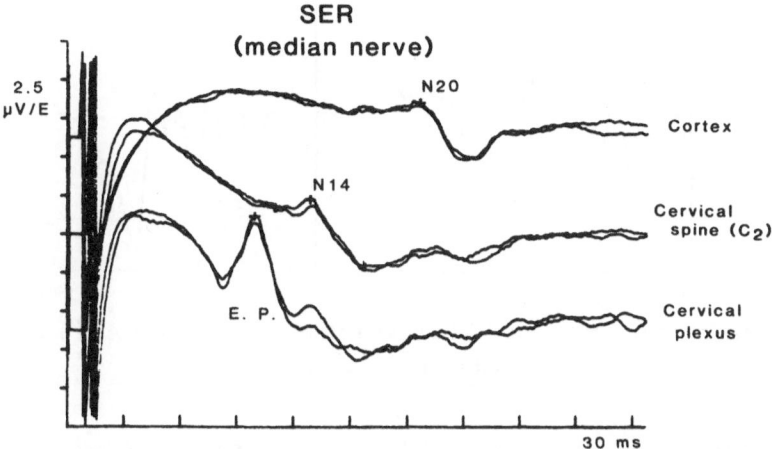

Fig. 1a. Somatosensory evoked response (SER, SEP): Lower tracings from Erb's point reveal a peak giving proof of correct stimulation. At approximately 13 to 14 ms the volley reaches the cervical spine at segment C_2 and after approximately 20 ms (N20) it reaches the scalp electrode

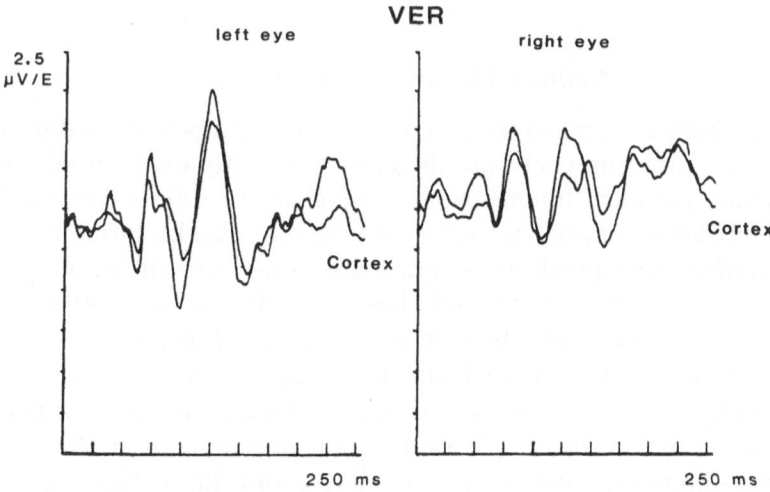

Fig. 1b. Visual evoked response (VER, VEP) as recorded from occipital scalp electrodes (C_z to O_z). A prominent positive peak (positive deflection down) is distinguished approximately after 100 ms (P100)

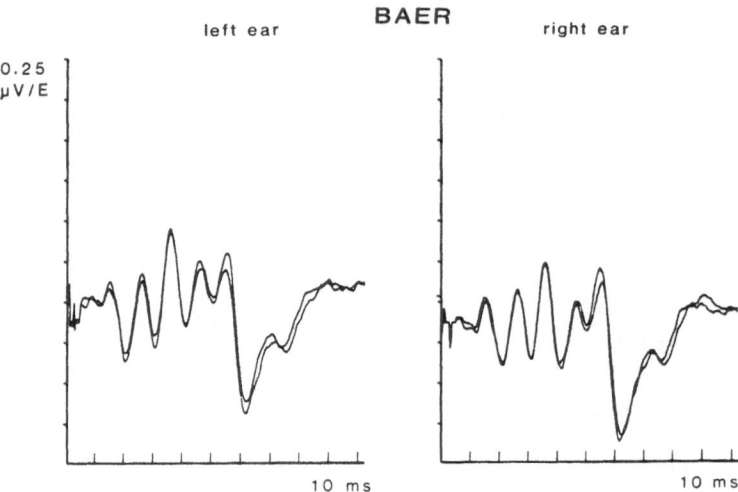

Fig. 1c. Brain stem (early) auditory evoked response (BAER, EAEP) as recorded from ipsilateral mastoid (C_z to $A_{1/2}$)

components are generated by which structure of the visual pathway. The usual time interval analysed after the stimulus is 200 ms to 300 ms. After stimulation of one eye only, the loss of the VER indicates a lesion distal to the chiasm. Bilateral loss of the VER with a bilaterally intact electroretinogram indicates either a severe global supratentorial disorder or a bilateral lesion frontal to the chiasm or of the chiasm itself.

Auditory Evoked Responses (BAER)

The techniques applied to elicit auditory evoked potentials are variable. Tone bursts, pips and clicks are the most frequently used stimuli. In brain stem auditory evoked potentials (BAER), clicks of 100 to 200 μs delivered monaurally are most popular. Analysis time may also be variable: 10 ms for short latency AER (also called brain stem auditory or far field evoked potentials), 10 to 60 ms for middle latency AER, and over 60 ms for long latency AER. Within the first 10 ms normally 7 prominent waves are distinguished (Fig. 1c). Wave I and II are supposed to be generated extracerebrally. Wave V is thought to be generated by the inferior colliculus, or by structures near by (Stockard and Rossiter 1977). An increased inter-peak-latency up to wave V will indicate some kind of functional lesion of the auditory pathway distal to the inferior colliculus. The generator of later components is uncertain (Halliday 1982, Stöhr et al. 1982). A normal BAER will not supply information about supratentorial functions.

Prognostic Value of Evoked Potentials in Severe Head Injury

The prognosis of a patient comatose due to a head injury is difficult to assess immediately after the trauma. A great number of publications have dealt with this problem. Frowein (1990) showed in over 2000 patients a correlation of coma grade, duration of coma, and age to survival and quality of survival. Braakman *et al.* (1980) have reported on the increasing accuracy of prediction based on clinical findings with time after the injury.

As the comatose patient cannot make known, what he feels, sees or hears, objective data on the function of somatosensory, visual and auditory pathways is desirable.

The first reports on evoked potentials from comatose patients came from Jouvet (1959), Bergamasco *et al.* (1966), Lille *et al.* (1967), Arfel (1967) and Götte *et al.* (1973). Greenberg *et al.* (1977) were the first to report on multimodality evoked potentials in a larger number of comatose patients after head injury. Results from the literature should be evaluated carefully: The term 'coma' may be used differently, the Glasgow Coma Scale being the most popular grading. Other differences may include variable technical standards for the measurement of evoked potentials. Most authors use a different grading for abnormal evoked potentials. Earlier studies on evoked potentials in comatose patients after head injury did not analyse the central conduction time of the SER. Several studies on the prognostic value of evoked potentials in comatose patients also included patients with a non-traumatic cause of coma.

Subsequently the role of evoked potentials in head injury will be discussed starting with a presentation of our experiences, which is followed by a discussion and review of the literature.

Patients and Methods

Seventy six comatose patients were examined. The ages ranged from 1.5 to 90 years. All patients had sustained a head injury.

Coma was graded into 4 grades according to Frowein (1976) (see also chapter by Frowein in this volume).

For technical data on evoked potentials see below: Chapter "How to do it".

The findings of evoked potentials were graded into 3 groups:

Grade I: Bilaterally reproducible responses with normal latencies (of N14, N20 and CCT for SER, and wave V and interpeak latency wave III to V for BAER).

Grade II: Unilaterally or bilaterally abnormal latencies and/or unilaterally no reproducible response.

Grade III: Bilaterally no reproducible response (in BAER bilaterally no response after wave II).

Statistical analysis was applied to the possible correlation among such variables as age, coma grade, duration of coma, SER, BAER and outcome (survival or non-survival).

Results

SER
Survival and non-survival as related to coma grade and SER are listed in Fig. 2a. Six out of 23 patients with normal SER (grade I) died. They had secondary deterioration not directly related to the courses of the initial coma (cardiovascular disorder, gastro-intestinal haemorrhage, alcoholism, recurrent sub-arachnoid haemorrhage).

With SER grade II, mortality amounted to 34%, and to 93% with SER grade III. The 2 survivors with bilateral absence of SER had unusual clinical courses: One was a 1.5 year-old-male with head injury. 4 serial investigations confirmed the absence of SER. 3 months later he survived in a vegetative state. The second patient was a 5 year-old-male with a cervical fracture and clinically complete transverse section at C4/5 combined with a head injury. After regaining consciousness the SER remained absent.

VER
The VER was recorded in 61 patients (see Fig. 2b). Fourteen out of 41 patients with a normal VER (grade I) died. With a VER grade II mortality reached 69%. One of 14 patients with a complete bilateral loss of the VER survived. This patient was admitted in coma grade I. When he regained consciousness he was found to be blind. Subsequently he also exhibited pituitary malfunction with diabetes insipidus.

BAER
In 76 comatose patients the BAER was elicited (see Fig. 2c). 8 out of 31 patients with normal BAER died. With a BAER grade II, 19 out of 30 patients died. With no reproducible BAER except for wave I and II, 13 out of 14 patients died. The only survivor was a 78 year-old male with a head injury, who had been deaf prior to the injury; this became known after he regained consciousness.

Discussion

After the first study on multimodality evoked potentials by Greenberg *et al.* in 1977, many studies followed (see Table 1). Also numerous investigations on single evoked response modalities, either SER or AER on comatose patients were reported (see Table 2, 3). It may be difficult to compare the prognostic value of evoked responses in these studies, as technical standards and the evaluation of evoked responses as well as the clinical

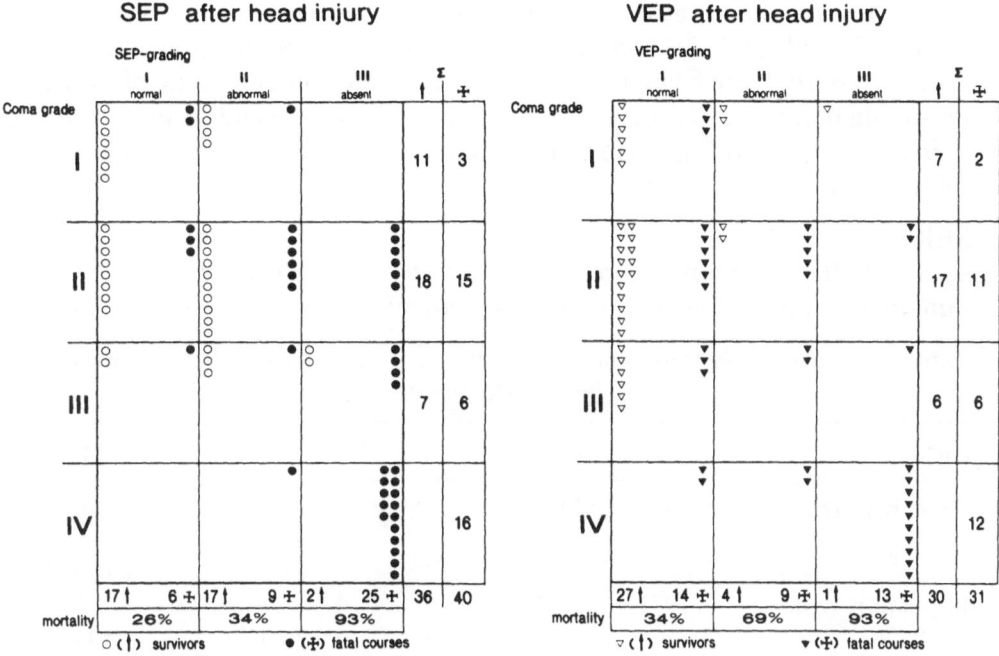

Fig. 2a. Relation of SER to coma
grade and outcome

Fig. 2b. Relation of VER to coma
grade and outcome

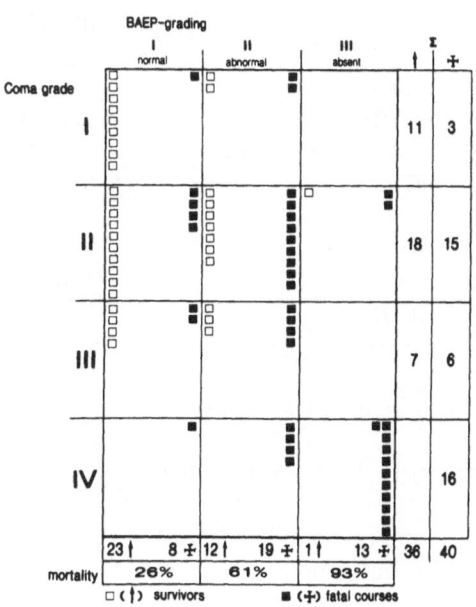

Fig. 2c. Relation of BAER to coma
grade and outcome

assessment of coma varied widely. Some of these studies (marked with asterisk in Tables 1, 2, 3) were carried out on head injury patients only. Others reported on head injuries and numerous other causes of coma, reasoning that prognosis of a comatose patient is less dependent on how the lesion was acquired (*e.g.* head injury, spontaneous haemorrhage etc.), than on its location and extent.

SER

From the current literature, SER proved to be significantly correlated to outcome in more than 400 comatose patients, most of them with head

Table 1. *Review of the Literature. Early Prognosis of Coma with Multimodality Evoked Responses*

Author	Year	n	SER	p	VER	p	AER	p
Greenberg *et al.*	1977	51*	l	+	l	+	s	nc
							1	+
								(< 0.05)
Rappaport *et al.*	1977	35			l	+	s	+
						(0.025)		(0.25)
							m	+
								(0.05)
							l	nc
Greenberg *et al.*	1981	100*	s	+	l	+	s	+
Lindsay *et al.*	1981	32*	l	+	l	+	s	nc
				(< 0.001)			1	+
								(0.005)
Mauguiere *et al.*	1982	20	s	+			s	+
Lütschg *et al.*	1983	43	s	+			m	+
				(< 0.001)				(0.001)
Anderson *et al.*	1984	39*	s	+	l	nc	s	+
				(0.0001)				(0.018)
Pfurtscheller *et al.*	1985	30	l	+		nc		
				(0.01)				
Cant *et al.*	1986	35	s	+			s	nc
				(0.01)				

Latency SER:
s – short (to 40 ms), m – middle (40 bis 70 ms) l – long (> 70 ms)
Latency AER:
s – short (up to 10 ms) m – middle (10–70 ms) l – long (over 70 ms)
nc not correlated.
p Correlation of evoked response with outcome (probability factor).
 + Good correlation of evoked response with outcome.
* Head injury only.

Table 2. *Review of the Literature: Early Prognosis of Coma with Somatosensory Evoked Responses Only*

Author	Year	Number	SER	p
De La Torre *et al.*	1978	17*	l	+
Hume *et al.*	1979	24*	s	+
				(p:0.03–0.001)
Rumpl *et al.*	1983	44*	s	+
				(p < 0.001)
Walser *et al.*	1986	63	s	+

SER Somatosensory evoked responses.
l Long latency (over 70 ms).
s Short latency.
p (Probability factor) correlation of SER with outcome.
+ SER and outcome are correlated.
* Head injury only.

Table 3. *Review of the Literature. Early Prognosis of Coma with Auditory Evoked Responses Only*

Author	Year	n	AER	p
Uziel and Benezech	1978	20	s	+
Seales *et al.*	1979	17*	s	+
Tsubokawa *et al.*	1980	64*	s	+
Karnaze *et al.*	1982	26*	s, l	+ (p < 0.01)
Hall *et al.*	1983	23*	s	(+)
Mjoen *et al.*	1983	11*	s	+
Zuccarello *et al.*	1983	20*	s	+
Rosenberg *et al.*	1984	25	s, m, l	not correlated
Facco *et al.*	1985	40*	s	+
Kaga *et al.*	1985	54	s, m, l	+
Karnaze *et al.*	1985	45*	s, l	+ (p < 0.0004)
Papanicolaou *et al.*	1986	38*	s	+

AER Auditory evoked responses.
l Long latency (over 70 ms).
m Middle latency (10–70 ms).
s Short latency (up to 10 ms).
p Correlation of AER with outcome (probability factor).
+ AER well correlated with outcome.
* Head injury only.

injury (see Tables 1, 2). In our patients SER proved to be of higher prognostic value in the prediction of a fatal outcome than in the prediction of a favorable outcome. This is consistent with everyday experience as many comatose patients die of intercurrent non related disorders (*e.g.* cardiac or respiratory failure) after the improvement of neurological deficits. Excluding secondary unrelated deterioration, Greenberg *et al.* (1981) found the accuracy of predicting outcome based on SER approaching 100%, which is similar to the study by Narayan et al. 1981.

The bilateral loss of SER indicates a severe functional lesion, as only 2 children survived this condition in our series. Zegers de Beyl and Brunko (1986) reported exceptions to this general rule, that the bilateral loss of SER indicates a poor outcome: They observed in 2 children and 2 adults after head injury not only survival and a favorable outcome, but the lost SER gradually reappeared in repeated recordings. These observations, however, seem to be very rare. As the somatosensory pathway runs through the entire length of the neuraxis from the upper cervical cord to the cortex, a normal SER is evidence of a normal infratentorial *and* supratentorial function of the somatosensory pathway. This may be the reason why in most studies the SER is more closely related to outcome than the BAER, as the BAER is generated infratentorially only and the VER is generated by supratentorial structures only.

VER

The prognostic value of the VER has attracted little attention, as it can supply information from supratentorial structures only. In comatose patients, however, the functions of the brain stem are of primary interest to predict survival or non-survival (Greenberg *et al.* 1977). As the VER is lost in most severe supratentorial lesions, loss of the VER after head injury usually indicates a fatal outcome, while a preserved VER by itself is of no prognostic value as it may be present in fatal infratentorial lesions. By contrast, Greenberg *et al.* 1977, Lindsay *et al.* 1981 and Rappaport *et al.* 1977 found a significant correlation of the VER with outcome, while Anderson *et al.* 1984 and Pfurtscheller *et al.* 1985 found no correlation.

The *diagnostic* value, however, was demonstrated in one of our patients: Bilateral loss of the VER in one comatose patient anticipated a visual defect which was confirmed when the patient regained consciousness. In suitable cases, the VER may possibly contribute to the decision to decompress the optic nerve after head injury (Dorfman *et al.* 1987).

BAER

Up to now, there have been more studies on the BAER and outcome in comatose patients, than on the SER. Anderson *et al.* 1984, Karnaze *et al.* 1982, 1985 and Tsubokawa *et al.* 1980 found a highly significant correlation of the BAER to outcome. Greenberg *et al.* (1977) reported no significant correlation (see Table 1). Similar to the SER the BAER seems to be a more accurate prognosticator of a fatal outcome, as we encountered only one patient with bilateral loss of the BAER, who survived with a peripheral hearing loss.

Conclusion

For the assessment of outcome of comatose patients after head injury on clinical manifestations only, the duration of coma is of fundamental significance (Frowein and Firsching 1990). Prediction of survival or non-survival will be more accurate the longer the interval after head injury. By contrast, a study of evoked potentials may be performed shortly after admission of the patient, making an early prognosis available. Somatosensory evoked responses, and to a lesser degree also brain stem auditory evoked responses, appear to increase the accuracy of prognosis significantly. The bilateral loss of the SER is a strong prognosticator of a fatal or poor outcome. The reappearance, however, of a bilaterally lost SER has been reported (Zegers de Beyl and Brunko 1986). Although this phenomenon seems to be a rare occurrence, it appears to be possible. The cost effectiveness of evoked potentials has been questioned by Kimura (1985) and by Lindsay *et al.* (1981), but most authors agree on the usefulness of recording evoked potentials in comatose patients after head injury.

Auditory Evoked Responses in Non-severe Head Injury

Several authors have commented on the short latency auditory evoked responses in minor head injury with a loss of consciousness of less than 24 hours. In earlier reports (Geets and Louette 1983, Noseworthy *et al.* 1981, Rowe and Carlson 1980) some relation was found between a post-concussion syndrome (usually referring to dizziness and slight mental disorders) and altered short latency auditory responses. Schoenhuber and Gentilini (1986) examined 30 patients. They found no correlation of 'subclinical brain stem involvement' as shown by abnormalities of auditory brain stem responses and impaired mental function or symptoms of the post-concussion syndrome. The value of auditory evoked responses in the assessment of minor head injury thus appears to be controversial.

Brain Death

Donors for organ transplantation are predominantly recruited from patients with fatal head injuries. From several countries there are reports

Table 4. *Review of the Literature. Evoked Responses in Brain Death*

Author	Patient a)	b)	SER	BAER
Anziska, Cracco 1980	11	7	+	
Goldie et al. 1981	35	5	+	+
Haupt 1985	20		(7 +)	+
Shiogai et al. 1989	108	27		+
				47 + +
Starr 1976	27			+
Stöhr et al. 1986	49			+
Trojaborg, Jørgensen 1973	31		+	

a) Primary and secondary brain lesion.
b) Primary supratentorial brain lesion only.
+ Single recording.
+ + Serial recordings, documentation of abolition.

on evoked potentials in brain death. In West Germany in 1986, documentation of the gradual loss of short latency auditory evoked responses has been used as a criterion in the diagnosis of brain death (Wissenschaftlicher Beirat 1986).

So far, there is a limited number of publications on the value of evoked potentials in brain death (see Table 4). International guidelines for the determination of brain death vary widely (Ad hoc committee, Harvard (1968), Allen et al. (1978), Pendl (1986), Walker (1985)). A common denominator are prerequisites which include a primary brain lesion and the exclusion of intoxication, hypothermia, endocrine or metabolic coma. Head injury, tumors, haemorrhages and acute hydrocephalus are considered primary brain lesions. Cerebral infarction due to transient cardiac arrest is regarded as a secondary brain lesion.

In West Germany brain death may be declared when the irreversibility of the loss of all integrative functions of the entire brain (brain stem and cortex) has been demonstrated. For this, apnoic brain stem areflexia must be demonstrated and then either a waiting period or additional tests are mandatory. Infratentorial primary brain lesions, however, may cause coma, cranial nerve areflexia and apnea, while the EEG may undoubtedly still detect intracerebral bioelectrical activity for up to 56 hours (Ferbert et al. 1985, 1986, Frowein et al. 1985). In West Germany brain death may not be declared as long as the EEG shows some EEG activity, because the loss of *all* brain functions is required. Therefore an EEG is mandatory in all primary infratentorial lesions. The proceedings thus depend on the differentiation of supratentorial from infratentorial lesions. Subsequently we will present a series of patients developing brain death and monitoring of evoked potentials.

Patients and Methods

Serial investigations demonstrating the abolition of evoked responses (see Fig. 3) were observed in 6 patients with supratentorial and in 4 patients with infratentorial brain lesions.

Criteria for apnoic brain stem areflexia were coma, absence of cranial nerve reflexes and apnea as confirmed by the apnea test. The cranial nerve reflexes tested were:

1. light reflexes
2. corneal reflexes
3. oculocephalic reflex
4. trigeminal nerve reflexes to painful stimuli
5. gag reflex
6. cough reflex

For the apnea test the patient was disconnected from the ventilator. 100% oxygen was administered via an intratracheal tube. Apnea was documented when no breathing could be detected after at least 10 min and blood gas analysis showed increased pCO_2 of at least 60 mmHg. In 2 patients with supratentorial lesions no blood gases were analysed. The technical standards and the grading of evoked potentials in this study on brain death were identical with the standards and gradings used in the study of comatose patients as stated above. The term "abolition" is used, when the loss of an evoked response is recorded after documentation of an initial at least partly preserved evoked response.

Supratentorial Lesions

The abolition of evoked responses in supratentorial brain lesions was documented in 6 cases. Of these, in 2 cases the apnea test was not carried out. Results are listed in Table 5.

SER

Cortical SER were investigated in all 6 patients. In 4 patients, the SER was already absent at the initial investigation. In 2 patients the bilateral abolition of the SER could be demonstrated.

VER

VERs were investigated in 3 patients. In all 3 patients the abolition of the VER was documented.

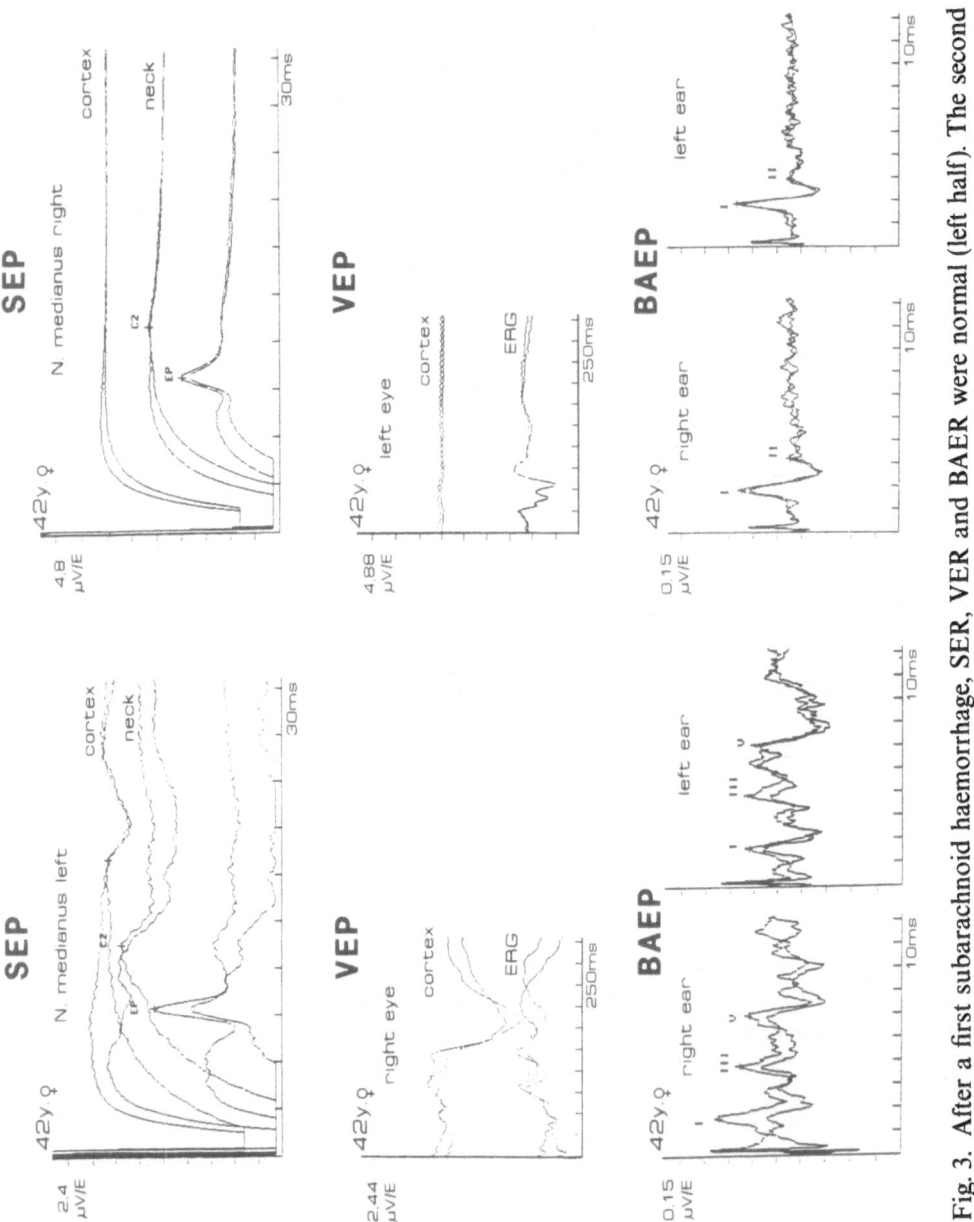

Fig. 3. After a first subarachnoid haemorrhage, SER, VER and BAER were normal (left half). The second haemorrhage proved to be fatal. Cortical components of any evoked potential were missing, while the cervical component of the SER, the electroretinogram (ERG) and wave I and II were preserved (right half)

Table 5. *Abolition of Evoked Potentials in Supratentorial Lesions* (n = 6)

No.	Diagnosis	Age	Coma	t	SER	VER	BAER	EEG-activity	Apnea-test (pCO$_2$mmHg)
39	HI	39	III	IR	III!̲		II	ES – 1 hr	
			BDS	09:00			III		78.6 + 1 hr
60	SAH	42	II	IR	I	I	I		
			IV	17:00	III!̲	III	III:̲	ES – 1 hr	
57	SAH	45	III	IR	III!:̲	II	I		
			IV	09:00	III!:̲		II		
			BDS	12:00	III	III	III		59 + 1 hr
55	HI	66	II	IR	I		II	δ – 1 hr	
			BDS	24:00	III!̲	III	III	ES – 1 hr	75 + 1 hr
35	intrac.h.	72	III	IR	III!̲	I	II	ES – 1 hr	
			BDS	16:00	III!̲	III	III.		76 + 1 hr
59	HI	80	III	IR	III!̲		II	ES – 1 hr	
			IV	08:00			III.		

Coma grade and evoked response grading in roman figures. No Patient number. IR Initial recording of evoked responses. t Time of recording, hr after IR. ES Electrocerebral silence of EEG. BDS Brain death syndrome. SAH Subarachnoid haemorrhage. HI Head injury. intrac.h. Intracerebral haematoma. SER, VER, BAER – somatosensory, visual, brain stem auditory evoked responses. !/∶ SER: wave N15 preserved unilaterally/bilaterally. ∶/∶ BAER: wave II preserved unilaterally/bilaterally. +/− Hours after/prior to recording of evoked responses. hr Hours.

BAER

BAERs were elicited in all 6 patients. In all patients, the initial BAER was preserved beyond wave III. In all patients the abolition of the BAER was observed. In apnoic brain stem areflexia only 1 patient had a unilaterally preserved wave I, in coma grade IV, one patient exhibited a bilaterally preserved wave II, another 80-year-old patient exhibited a unilaterally preserved wave I.

Infratentorial Lesions

In 4 patients serial investigations of multimodality evoked potentials were carried out (see Table 6). Subsequently the clinical course of each of these cases will be presented, as these infratentorial lesions usually show a sequence of symptoms different from supratentorial lesions.

Case 1

A 17-year-old female (71–87) suffered a ruptured liver after being hit by a horse. After the extensive loss of blood, a critical drop of the blood pressure to 50 mmHg was recorded during operative repair. Postoperatively she remained comatose. CT showed a large ischemic lesion of the right cerebellar hemisphere. 24 hours later EEG showed α activity. While the SER was lost bilaterally, VER and BAER were normal. 72 hours later EEG showed electrocerebral silence. At this point the BAER was also lost bilaterally. In this case, the loss of supratentorial functions coincided with the loss of infratentorial functions.

Case 2

A 33-year-old female (1625–86) remained comatose after removal of a clivus meningioma. EEG revealed δ-activity 2 days after operation. While the cervical SER was preserved, there was no cortical SER. The VER was partly preserved, the BAER was normal. Another 21 hours later, EEG showed electrocerebral silence and the VER and the BAER were lost.

Case 3

This 45-year-old male (1181–86) had a pontine haemorrhage and was admitted in coma grade III. The SER was lost bilaterally, VER was normal, the BAER was partly preserved. 33 hours later, EEG showed electrocerebral silence and the BAER was lost. Immediately afterwards apnea was confirmed by the apnea test.

Case 4

This 71-year-old female (1453–86) on anticoagulation suffered a pontine haemorrhage. Upon admission coma grade III was noted. EEG showed α, β and δ activity. 3 hours after admission, VER was preserved, while the SER

Table 6. *Abolition of Evoked Potentials in Infratentorial Lesions (n = 4)*

No.	Diagnosis	Age	Coma	t	SER	VER	BAER	EEG-activity	Apnea-test (pCO_2 mmHg)
67	cereb. infarct.	17	II	IR	III!⁻	I	I	ES − 2 hr	$-\frac{1}{2}$ hr
			BDS	48:00					
63	postop.c. infratent. tumor	35	III	IR	III!⁻	II	I	δ − 4 hr	38.8 − $\frac{1}{2}$ hr
			BDS	21:00	III	III	III	N − 1 hr	(50 min)
38	pontine h.	45	II	IR	III	I	II	δ − 1 hr	
			II	10:00		III	II	ES − $\frac{1}{2}$ hr	
			BDS	34:00			III		
50	pontine h.	71	III	IR	III!⁻	I	III	αβδ − 3 hr	73 + $\frac{1}{4}$ hr
			BDS	15:00		III	III	αβδ − 1 hr	67.3 + 1 hr
									71 + 2 hr

Coma grade and evoked response grading in roman figures. No Patient number. IR Initial recording of evoked responses. t Time of recording, hr after IR. ES Electrocerebral silence of EEG. BDS Brain death syndrome. cereb. infarct. Cerebellar infarction. postop. c. infratent. tumor postoperative course infratentorial tumor. h. Haemorrhage. SER, VER, BAER Somatosensory, visual, brain stem auditory evoked responses. !/⁻ SER: wave N15 preserved unilaterally/bilaterally. ./⁻ BAER: wave I preserved unilaterally/bilaterally. :/⁻ BAER: wave II preserved unilaterally/bilaterally. + / − Hours after/prior to recording of evoked responses. hr Hours.

and BAER were lost bilaterally. One hour later apnea was confirmed by apnea test. Another 17 hours later EEG still showed some α- and δ-activity. Another apnea test immediately afterwards revealed loss of spontaneous respiration. The patient died shortly afterwards of cardiac arrest.

While in case 1, 2 and 3 the loss of supratentorial and infratentorial functions was noted simultaneously, coma, cranial nerve areflexia and apnea were noted in case 4 while the EEG still detected some cortical functions.

Discussion

No matter where the brain lesion was located, all patients exhibiting apnoic brain stem areflexia proved to have a loss of SER, VER and BAER (except wave I). In all serial investigations, the reappearance of any evoked response was never observed, once it had been completely lost. The clinical course of events leading to brain death in supratentorial lesions is usually different from infratentorial lesions:

In supratentorial lesions, electrocerebral silence of the EEG may be observed prior to the loss of spontaneous respiration. Frowein *et al.* (1985) observed electrocerebral silence with preserved respiration for 10 hours in a 14 year old girl after a bicycle accident. In these primary supratentorial lesions, once coma, cranial nerve areflexia and apnea is noted, there are 4 confirmatory tests in W. Germany, each recommended as alternatives, to prove the irreversibility of this loss of all integrative brain functions: 1st a waiting period of 12 hours with serial confirmation of brain death syndrome, 2nd: a 30 minute EEG displaying electrocerebral silence, 3rd: abolition of BAER (for definition of "abolition", see above), 4th: angiographic proof of arrest of cerebral perfusion. This successive order of diagnostic steps is mandatory and may not be reversed. The confirmatory tests may only contribute to the final declaration of brain death, *after* coma, cranial nerve areflexia and apnea has been noted. Brain death may not be declared when electrocerebral silence coincides with preserved respiration. The role of evoked responses depends on the modality investigated. Consistent with other authors (Anziska and Cracco 1980, Ganes and Nakstad 1984, Goldie *et al.* 1981, Starr 1976), cortical *SERs* were lost in all our patients when apnea was found. However, survival with bilateral loss of cortical SER was seen in 2 children from our series. It has also been reported by Zegers de Beyl and Brunko (1986) in adults. Survival after development of apnoic brain stem areflexia and bilateral abolition of the cortical SER has not been reported yet.

A preserved retinal response in brain death has been known for a long time (Wilkus *et al.* 1971). While a completely preserved *VER* is not compatible with brain death, the loss of VER has not been used as confirmatory test, as the generator of later components of the VER is uncertain. In one of our patients bilateral loss of VER was observed with preserved α-activity of the EEG.

The single recording of the absence of the *BAER* has been reported repeatedly in brain death (see Table 4). Serial investigation with documentation of the stepwise abolition of BAER have been reported in a few studies only (Buchner *et al.* 1989, Ferbert *et al.* 1987, Klug 1982, Shiogai *et al.* 1989, Tapie *et al.* 1985). Consistent with other authors (Goldie *et al.* 1981, Hall *et al.* 1985, Shiogai *et al.* 1989, Starr 1977, Stöhr *et al.* 1986, Tapie *et al.* 1985, Trojaborg and Jørgensen 1973) wave I was preserved in a few cases only (25%), mostly no reproducible waves were distinguished. Wave II has been considered compatible with brain death (Guerit and Mahieu 1986) as it is thought to be generated extracerebrally (Stockard *et al.* 1977). As the loss of brain stem function is usually secondary to the loss of cortical functions during the development of brain death after primary supratentorial lesions, the abolition of BAER has been adopted as a confirmatory test for the declaration of brain death. It must be remembered, however, this is only valid after confirmed primary supratentorial lesions, and *serial* investigations demonstrating the abolition of BAER are mandatory. If the kind or location of brain lesion is uncertain, the abolition of BAER should therefore not be used as a confirmatory test.

In *infratentorial* lesions some cortical function may be preserved for a variable interval after the manifestation of coma, apnea and cranial nerve areflexia (see Table 7). EEG currently seems to be most appropriate to pick up the final loss of cortical function in these patients. The recording of the abolition of BAER in infratentorial lesions will not give information about supratentorial functions and is therefore not a relevant confirmatory test. At the most it may confirm the loss of brain stem function in questionable drug intoxication (barbiturates), as the BAER is particularly resistant to drugs (Clar and Rosner 1973, Guerit 1986, Sutton *et al.* 1982).

Reliability of Evoked Potentials

In any medical test, the risk of false findings may not be excluded with absolute certainty. *False positive* results in evoked potentials are the recording of potentials, that do not exist in reality. This is highly unlikely to happen if correct techniques include a second test to demonstrate reproducibility of the potentials found. The probability, that unrelated artifacts add up to the same evoked response pattern in 2 averaged series is extremely small. In doubtful cases, muscle relaxants may be used (Haupt 1986). *False negative* findings are the absence of reproducible responses, whereas in reality there are at least some evoked responses preserved. There may be

Table 7. *Review of the Literature* (from Frowein *et al.* 1987). *Development of Brain Death Syndrome After Primary Infratentorial Brain Lesions*

Author	Year	Diagnosis	SER	VER	BAER	EEG	After apnea
Frowein *et al.*	1985	cerebell. h.	0		0	θ	< 22 hr
Ferbert *et al.*	1986	cerebell. + pontine h.	0	+	0	α	> 56 hr
Rodin *et al.*	1985	postop. c. cbpa-tumor	0	+	0	$\alpha\theta\delta$	14 days
Ferbert *et al.*	1985	bas. art. thr.	0		0		3 hr
Ferbert *et al.*	1986	bas. art. thr.	0	+	0	θ	49 hr
Ferbert *et al.*	1986	bas. art. thr.	0		0	α	2.5 hr
Haupt	1986	bas. art. thr.	0	+	+	$\theta\delta$	> 24 hr

SER, VER, BAER – somatosensory, visual, brain stem auditory evoked response.
cerebell. Cerebellar.
h. Haematoma.
postop. c. cbpa Postoperative course cerebello-pontine-angle tumor.
bas. art. thr. Basilar artery thrombosis.
 + Preserved.
0 Not preserved.

two reasons for this: the investigator and/or the equipment. Only qualified personnel should record evoked potentials, as there are many possibilities for improper usage. The implications especially in the diagnosis of brain death are serious. Faulty equipment cannot be excluded, the function of the stimulator and impedances of electrodes must therefore be checked prior to each recording.

The *reappearance* of bilaterally lost BAERs has been observed in 2 instances (Rossini *et al.* 1982, Taylor *et al.* 1983). In these cases, however, the underlying cause was not a primary brain lesion (near drowning, transient cardiac arrest) and there were no clinical signs of brain death. There are no reports of reappearance of bilaterally lost BAER after manifest clinical apnoic brainstem areflexia.

Conclusion

After registration of coma, cranial nerve areflexia and apnea there are several confirmatory tests in use internationally for the declaration of brain death. The stepwise abolition of BAER can be considered a reliable confirmatory test after primary supratentorial brain lesions. The advantage of the BAER over other tests is its relative resistance to drug intoxication. It is therefore a valuable additional tool in the diagnosis of brain death.

How to Do It

There are several ways and various pieces of equipment to record evoked potentials. Unfortunately there is no agreement on uniform standards internationally. The investigator should have experience with evoked potentials and be familiar with the equipment. Subsequently the data of the technical parameters we found useful are presented:

SER

For the SER, preferably the median nerve was stimulated at the wrist. The stimulus was increased until a twitch of the thumb was seen (maximum 20 mA). If too many artifacts made a muscle relaxant necessary, the stimulating electrodes were checked before the relaxant was given.

stimulus:	200 μs < 20 mA at 5, 4 Hz, 2 × 512 runs
bandpass:	LF:HF — 30:3000 Hz
analysis time:	30 ms
montage:	1. Erb's point — Fz, 2. neck at C_2 — Fz, 3. contralateral scalp (C_3 or C_4) — Fz

VER

As the comatose patient cannot open his eyes, goggles are necessary to elicit the VER.

stimulus:	unilateral flash ($\approx$ 1 ms) 2 × 256 at 1.7 Hz
bandpass:	LF:HF — 1:100 Hz
analysis time:	250 ms
montage:	1. ipsilateral lateral eyelid — Cz, 2. Oz — Cz

BAER

Headphones are advisable to elicit auditory evoked potentials in comatose patients.

stimulus:	2 × 1024 unilateral alternating rarefaction/condensation 200 μs clicks delivered at 11.1 Hz with 95 dB and 65 dB contralateral white noise masking
bandpass:	LF:HF — 150:3000 Hz
analysis time:	10 ms
montage:	ipsilateral ear lobe — Cz

References

1. Ad hoc committee of the Harvard Medical School to examine the definition of brain death: A definition of irreversible coma. JAMA 205, 337–340, 1968

2. Allen N, Burkholder J, Comiscioni J (1978) Clinical criteria of brain death. In: Korein J (ed) Brain death. Interrelated medical and social issues. The New York Academy of Sciences, New York

3. Anderson D, Bundlie S, Rockswold G (1984) Multimodality evoked potentials in closed head trauma. Arch Neurol 41: 369–374

4. Anziska B, Cracco R (1980) Short latency somatosensory evoked potentials in brain dead patients. Arch Neurol 37: 222–225

5. Arfel G (1967) Stimulations visuelles et silence cerebral. Electroencephalogr Clin Neurophysiol 23: 172–175

6. Bergamasco B, Bergamini L, Mombelli A, Mutani R (1966) Longitudinal study of visual evoked potentials in subjects in posttraumatic coma. Schw Arch Neurochir Psychiat 97: 1–10

7. Braakman R, Gelpke G, Habbema J, Maas A, Minderhoud J (1980) Systematic selection of prognostic features in patients with severe head injury. Neurosurgery 6: 362–370

8. Buchner H, Ferbert A, Hacke W (1989) Serial recording of median nerve stimulated somatosensory evoked potentials in brain death. Neurosurg Rev 12 [Suppl] 1: 348–352

9. Cant B, Hume A, Judson J, Shaw N (1986) The assessment of severe head injury by short-latency somatosensory and brain-stem auditory evoked potentials. Electroencephalogr Clin Neurophysiol 65: 188–195

10. Clar D, Rosner B (1973) Neurophysiological effects of general anaesthetics. 1. The electroencephalogram and sensory evoked potentials in man. Anesthesiology 38: 564–582

11. De La Torre J, Trimble J, Beard R, Hanlon K, Surgeon J (1978) Somatosensory evoked potentials for the prognosis of coma in humans. Exp Neurol 60: 304–317

12. Dorfman L, Gaynon M, Ceranski J, Louis A, Howard J (1987) Visual electrical evoked potentials: evaluation of ocular injuries. Neurology 37: 123–128

13. Facco E, Martini A, Zuccarello M, Agnoletto M, Giron G (1985) Is the auditory brain stem response (ABR) effective in the assessment of posttraumatic coma? Electroencephalogr Clin Neurophysiol 62: 332–337

14. Ferbert A, Buchner H, Ringelstein E, Hacke W (1985) Der Hirnstammtod bei Basilaristhrombose – eine besondere Variante des Hirntodes? In: Gänshirt H, Berlit P, Haack G (Hrsg) Kardiovaskuläre Erkrankungen und Nervensystem, Neurotoxikologie, Probleme des Hirntodes (Jahrestagung vom 19–22 Sept 1984 in Heidelberg). Springer, Berlin Heidelberg New York Tokyo

15. Ferbert A, Buchner H, Ringelstein E, Hacke W (1986) Isolated brain-stem death. Case report with demonstration of preserved visual evoked potentials. Electroencephalogr Clin Neurophysiol 65: 157–160

16. Ferbert A, Buchner H, Ringelstein E, Hacke W (1987) Brain death from infratentorial lesions: clinical, neurophysiological and transcranial Doppler

Ultrasound findings. Neurosurg Rev [Suppl 1]

17. Frowein RA (1976) Classification of coma. Acta Neurochir (Wien) 34: 5–10
18. Frowein RA, Firsching R (1990) Classification of head injury. In: Vinken PJ, Bruyn W (eds) Handbook of clinical neurology, Vol 59. Elsevier, North Holland Publ Comp, Amsterdam, in press
19. Frowein RA, Gänshirt H, Hamel E, Haupt WF, Firsching R (1987) Hirntod-Diagnostik bei primär infratentorieller Hirnschädigung. Nervenarzt 58: 165–170
20. Frowein RA, Richard KE, Hamel E (1985) Probleme des Hirntodes. In: Gänshirt H, Berlit P, Haack G (Hrsg) Kardiovaskuläre Erkrankungen und Nervensystem, Neurotoxikologie. Probleme des Hirntodes. (Jahrestagung vom 19–22 Sept 1984 in Heidelberg). Springer, Berlin Heidelberg New York Tokyo
21. Ganes T, Nakstad P (1984) Subcomponents of the cervical evoked response in patients with intracerebral circulatory arrest. J Neurol Neurosurg Psychiatry 47: 292–297
22. Geets W, Louette N (1983) EEG et potentiels evoques du tronc cerebral dans 125 commotions recentes. Rev EEG Neurophysiol 13: 253–258
23. Goldie W, Chiappa K, Young R, Brooks E (1981) Brainstem auditory and short latency somatosensory evoked responses in brain death. Neurol 31: 248–256
24. Götte J, Kubicki S, Kühn K, Stölzel R (1973) Klinische Anwendung somatosensorisch evozierter kortikaler Potentiale II. Untersuchungen an Patienten einer Reanimationsabteilung. Z EEG–EMG 4: 86–97
25. Greenberg R, Mayer D, Becker D, Miller J (1977) Evaluation of brain function in severe human head trauma with multimodality evoked potentials. J Neurosurg 47: 150–162
26. Greenberg R, Newlon P, Hyatt M, Narayan R, Becker D (1981) Prognostic implications of early multimodality evoked potentials in severely head injured patients. J Neurosurg 55: 227–236
27. Guerit J (1986) Unexpected myogenic contaminants observed in the somatosensory evoked potentials recorded in one brain dead patient. Electroencephalogr Clin Neurophysiol 64: 21–26
28. Guerit J, Mahieu P (1986) Are evoked potentials a valuable tool for the diagnosis of brain death? Transplantation Proceedings Vol XVIII, 3, 386–387
29. Hall J, Huangfu M, Gennarelli T, Kimmelman C, Dolinskas C (1983) Auditory brainstem abnormalities in experimental and clinical acute severe head injury. Transactions of the Penn Acad of Ophthalm and Otolaryng
30. Hall J, Mackey-Hagardine J, Kim E (1985) Auditory brain stem response in determination of brain death. Arch Otolaryng 111: 613–620
31. Halliday A (ed) (1982) Evoked potentials in clinical testing. Churchill Livingstone, Edinburgh London Melbourne New York
32. Haupt WF (1985) Diagnostische und prognostische Wertigkeit multimodaler evozierter Potentiale bei Hirnstammprozessen. Habilitationsschrift, Köln
33. Haupt WF (1986) Kraniale Muskelaktivität beim dissoziierten Hirntod. Nervenarzt 57: 145–148

34. Hume A, Cant B, Shaw N (1979) Central somatosensory conduction time in comatose patients. Ann Neurol 5: 379–384
35. Jouvet M (1959) Diagnostic electro-sous-corticographique de la mort du systeme nerveux central au cours de certains comas. Electroencephalogr Clin Neurophysiol 11: 805–808
36. Kaga K, Nagai T, Takamoti A, Matsch R (1985) Auditory short, middle and long latency responses in acutely comatose patients. Laryng 95: 321–325
37. Karnaze D, Marshall L, McCarthy C, Klauber M, Bickford R (1982) Localizing and prognostic value of auditory evoked responses in coma after closed head injury. Neurol 32: 299–302
38. Karnaze D, Weiner J, Marshall L (1985) Auditory evoked potentials in coma after closed head injury: A clinical-neurophysiologic coma scale for predicting outcome. Neurol 35: 1122–1126
39. Kimura J (1985) Abuse and misuse of evoked potentials as a diagnostic tool. Arch Neurol 42: 78–80
40. Klug N (1982) Brainstem auditory evoked potentials in syndromes of decerebration, the bulbar syndrome and in central death. J Neurol 227: 219–228
41. Lille F, Borlone M, Lerique A, Scherrer J, Thieffry S (1967) Evaluation de la profondeur du coma chez l'enfant par la technique des potentiels evoques. Rev Neurol 117: 216–217
42. Lindsay K, Carlin J, Kennedy I, Fry J, McInnes A, Teasdale G (1981) Evoked potentials in severe head injury – analysis and relation to outcome. J Neurol Neurosurg Psychiatry 44: 796–802
43. Lütschg J, Pfenninger J, Ludin H, Vasella F (1983) Brain-stem auditory evoked potentials and early somatosensory evoked potentials in neurointensively treated comatose children. Am J Dis Child 137: 421–426
44. Mauguiere F, Grand C, Fischer C, Courjon J (1982) Aspects des potentiels evoques auditifs et somesthesiques precoces dans les comas neurologiques et la mort cerebrale. Rev EEG Neurophysiol 12: 280–286
45. Mjoen S, Nordby H, Torvik A (1983) Auditory evoked brainstem responses (ABR) in coma due to severe head trauma. Acta Otolaryngol 95: 131–138
46. Narayan RK, Greenberg RP, Miller JD, Enas GG, Choin SC, Kishore PRS, Selhorst JB, Lutz HA, Becker DP (1981) Improved confidence of outcome prediction in severe head injury. J Neurosurg 54: 751–762
47. Noseworthy J, Miller J, Murray T, Regan D (1981) Auditory brainstem responses in postconcussion syndrome. Arch Neurol 38: 275–278
48. Papanicolaou A, Loring D, Eisenberg H, Raz N, Contreras F (1986) Auditory brain stem evoked responses in comatose head-injured patients. Neurosurg 18: 173–175
49. Pendl G (1986) Der Hirntod. Springer, Wien New York
50. Pfurtscheller G, Schwarz G, Gravenstein N (1985) Clinical relevance of long-latency SEPs and VEPs during coma and emergence of coma. Electroencephalogr Clin Neurophysiol 62: 88–98
51. Rappaport M, Hall K, Hopkins K, Belleza T, Berrol S, Reynolds G (1977) Evoked brain potentials and disability in brain-damaged patients. Arch Phys

Med Rehabil 85: 333–338

52. Rodin E, Tahir S, Austin O, Andaya L (1985) Brainstem death. Clin Electroencephalogr 16: 63–71

53. Rosenberg M, Wogensen K, Starr A (1984) Auditory brain-stem and middle- and long-latency evoked potentials in coma. Arch Neurol 41: 835–838

54. Rossini P, Kula R, House W, Cracco R (1982) Alteration of brainstem auditory evoked responses following cardiorespiratory arrest and resuscitation. Electroencephalogr Clin Neurophysiol 54: 232–234

55. Rowe M, Carlson C (1980) Brainstem auditory evoked potentials in postconcussion dizziness. Arch Neurol 37: 679–683

56. Rumpl E, Prugger M, Gerstenbrand F, Hackl J, Pallua A (1983) Central somatosensory conduction time and short latency somatosensory evoked potentials in posttraumatic coma. Electroencephalogr Clin Neurophysiol 56: 583–596

57. Schoenhuber R, Gentilini M (1986) Auditory brain stem responses in the prognosis of late postconcussional symptoms and neuropsychological dysfunction after minor head injury. Neurosurg 19: 532–534

58. Seales D, Rossiter V, Weinstein M (1979) Brainstem auditory evoked responses in patients comatose as a result of blunt head trauma. Trauma 19: 347–353

59. Shiogai T, Takeuchi K, Ogashiwa M, Hara M, Kadowaki C, Maeda T, Nakamura M (1989) Brainstem auditory evoked potential monitoring in the diagnosis of brain death. Neurosurg Rev 12 [Suppl 1]: 328–339

60. Starr A (1976) Auditory brainstem responses in brain death. Brain 99: 543–554

61. Stockard J, Rossiter V (1977) Clinical and pathologic correlates of brain stem auditory response abnormalities. Neurol 27: 316–325

62. Stockard JJ, Stockard JE, Sharbrough F (1977) Detection and localization of occult lesions with brainstem auditory responses. Mayo Clin Proc 52: 761–769

63. Stöhr M, Dichgans J, Diener H, Buettner U (1982) Evozierte Potentiale. Springer, Berlin Heidelberg New York

64. Stöhr M, Trost E, Ullrich A, Riffel B, Wengert P (1986) Bedeutung der frühen akustisch evozierten Potentiale bei der Feststellung des Hirntodes. Dtsch med Wschr 11: 1515–1519

65. Sutton L, Frewen T, Marsh R, Jaggi J, Bruce D (1982) The effects of deep barbiturate coma on multimodality evoked potentials. J Neurosurg 57: 178–185

66. Tapie P, Feblot P, Tuillas M, Lepetit J, Croguennec J (1985) Potentiels evoques auditifs precoces du tronc cerebral dans la mort cerebrale. Rev Electroencephalogr Clin Neurophysiol 14: 329–332

67. Taylor M, Houston B, Lowry N (1983) Recovery of auditory brain-stem responses after a severe hypoxic ischemic insult. Medical Intelligence 309: 1169–1170

68. Trojaborg W, Jørgensen E (1973) Evoked cortical potentials in patients with "isoelectric" EEGs. Electroencephalogr Clin Neurophysiol 35: 301–309

69. Tsubokawa T, Nishimoto H, Yamamoto T, Kitamura M, Katayama Y, Moriyasu N (1980) Assessment of brainstem damage by the auditory brainstem response in acute severe head injury. J Neurol Neurosurg Psychiatry 43: 1005–1011

70. Uziel A, Benezech J (1978) Auditory brainstem responses in comatose patients: Relationship with brainstem reflexes and levels of coma. Electroencephalogr Clin Neurophysiol 45: 515–524

71. Walker EA (1985) Cerebral death, 3rd ed. Urban und Schwarzenberg, Baltimore München

72. Walser H, Sutter M (1986) Koma. Thieme, Stuttgart New York

73. Wilkus R, Chatrian G, Lettich E (1971) The electroretinogram during terminal anoxia in humans. Electroencephalogr Clin Neurophysiol 31: 537–546

74. Wissenschaftlicher Beirat der Bundesärztekammer (1986) Kriterien des Hirntodes. Entscheidungshilfen zur Feststellung des Hirntodes. Deutsch Ärztebl 83: 2940–2946

75. Zegers de Beyl D, Brunko E (1986) Prediction of chronic vegetative state with somatosensory evoked potentials. Neurol 36: 134

76. Zuccarello M, Fiore D, Pardatscher K, Paolin A, Trincia G, Andrioli G (1983) Importance of auditory brainstem responses in the CT diagnosis of traumatic brainstem lesions. AJNR 4: 481–483

Subject Index

Thoracic and Lumbar Spine and Spinal Cord Injuries

Managing Editor: Phillip Harris

Editorial Board: J. C. Christensen, G. J. Dohrmann, S. El-Gindi, J. W. Glowacki, B. Ramamurthi

(Advances in Neurotraumatology, Volume 2)

1987. 64 figs. XIV, 211 pages.
Cloth DM 110,–, öS 770,–
ISBN 3-211-81928-2

Prices are subject to change without notice

Owing to their frequency and possible consequences, traumatic lesions of the thoraco-lumbar spine represent a special point of interest within the field of neurotraumatology. Traffic accidents are the commonest cause. According to previously published statistics nearly 50% of the cases affect the thoraco-lumbar junction.

After an introduction on epidemiology and the biomechanical properties, the clinical aspects of these injuries are discussed in detail. The radiological chapter stresses the generally available techniques as X-ray and CT scan and points out the future possibilities of MRI.

As regards treatment, there is still conflict between those in favor of conservative treatment, the indications of which are well known, and those advocating surgery at a fairly early stage. In all cases, the aim is to reduce and stabilize osseous lesions, lessen the risk of complications and facilitate rehabilitation. Chapters on revascularisation of the spinal cord and an appendix on head injuries complete the volume.

Springer-Verlag Wien New York

Extracerebral Collections

Managing Editor: Robert L. McLaurin

Editorial Board: P. Benedek, J. Brihaye, H. Makino, I. Oprescu, A. de Vasconcellos Marques

(Advances in Neurotraumatology, Volume 1)

1986. 72 figs. XI, 255 pages.
Cloth DM 112,–, öS 784,–
ISBN 3-211-81876-6

Prices are subject to change without notice

This volume is a collection of authoritative discussions of the various types of traumatic blood and fluid accumulations. Each chapter is designed to define in depth the pathogenesis, pathophysiology, diagnosis and treatment of one specific type of extracerebral collection. In addition, for those individuals who are principally interested in the practical aspects of recognition and management, the authors have included brief "How to do it" sections to their chapters. The material contained in each chapter includes the principles of clinical diagnosis when minimum adjunctive methods are available as well as the usage of sophisticated techniques if such can be employed.

In addition to chapters dealing with extracerebral collections, this volume contains chapters dealing with prognosis in severe craniocerebral trauma and with the principles, value, and methodology of intracranial pressure monitoring.

Contents: P. Guillermain: Traumatic Extradural Hematomas – D. P. Becker: Acute Subdural Hematomas – J. Brihaye: Chronic Subdural Hematoma – S. Matsumoto, N. Tamaki: Subdural Hydromas – M. Choux: Extracerebral Hematomas in Children – L. F. Marshall: Intracranial Pressure Monitoring: Theory and Practice – J. D. Miller: Prediction of Outcome After Head Injury – A Critical Review – R. L. McLaurin: Epilogue – Subject Index.

Springer-Verlag Wien New York